W9-CBV-917

Anatomy of a Rose

Other books by Sharman Apt Russell

Songs of the Fluteplayer

Kill the Cowboy

When the Land Was Young

Anatomy of a Rose

Exploring the Secret Life of Flowers

Sharman Apt Russell

ILLUSTRATIONS

Liddy Hubbell

PERSEUS PUBLISHING

Cambridge, Massachusetts

Many of the designations used by manufacturers and sellers to distinguish their products are claimed as trademarks. Where those designations appear in this book and Perseus Books was aware of a trademark claim, the designations have been printed in initial capital letters.

A CIP record is available from the Library of Congress
ISBN 0-7382-0208-8

Printed in the United States of America.
Perseus Publishing is a member of the Perseus Books Group

Text design by Elizabeth Lahey

Set in Fairfield 11/14 pt by Perseus Publishing Services
2 3 4 5 6 7 8 9 10–03 02 01
First printing, February 2001
Find Perseus Publishing on the World Wide Web at
http://www.perseuspublishing.com

To Peter ~ I love you

Contents

ILLUSTRATIONS

Acknowledgments

I want to thank my husband, Peter, and my children, David and Maria, for their constant and important support. Peter was my editor at home. Maria was a cheerful companion at the Sixteenth International Botanical Congress in St. Louis, Missouri. My friend Gail Stanford accompanied me to Huntington Gardens.

Amanda Cook, my editor at Perseus Books, was a tremendous help in shaping this book and in encouraging me to trust my instincts and personal voice.

In writing this manuscript, I sent out pleading letters and e-mails to perfect strangers, men and women whose research I was using and whose advice I needed. The following scientists responded generously with their time and efforts. I cannot thank them enough.

Despite the invaluable help of these people, any mistakes in the final manuscript are entirely my own.

Jack Carter gave me some good advice at the very start.

Nick Waser helped me from the beginning to the end, patiently reading through and commenting on many chapters. In some cases, he read through them twice. Nick has been an active researcher in floral biology for thirty years. I credit his work extensively in the selected bibliography and notes. His willingness to help reflects his genuine commitment to education and conservation. I also suspect that Nick is one of those tireless people who can juggle six balls in the air while standing on one foot and lecturing about delphiniums.

Lars Chittka is an extraordinary role model for anyone entering the field of science. His research on bee vision and insect behavior is simply exciting. My first draft of "The Blind Voyeur" included many more direct references to his work, some of which I edited out, in deference to my lay audience. Any reader who pursues the subject will enjoy seeing Lars's name in both popular and academic science magazines.

Rob Raguso was an important source of encouragement and help. Rob is the subject of "What We Don't Know" and my adviser for "Smelling Like a Rose."

Alison Brody provided me with a number of papers on her research, particularly on nectar robbing. She also commented on the beginning chapters.

Martha Weiss read over the text concerning her work on butterflies and changes in flower color. Her research is cited heavily in "The Physics of Beauty."

Steve MacDonald advised and fretted with me concerning how to discuss the complexity of evolution in a few short paragraphs.

Via the miracle of e-mail, Roger Seymour, a professor at the University of Adelaide in Australia, looked over his material for "In the Heat of the Night."

Judith Bronstein kindly read over and commented on "Dirty Tricks," a chapter that highlights her work on mutualism, yuccas, and yucca moths.

Brent Mishler, a spokesperson for the project Deep Green, read over "The Tower of Babel and the Tree of Life" and "Flowers and Dinosaurs."

Kirk Johnson also helped keep me on track in "Flowers and Dinosaurs."

Finally I must thank the staff of the Interlibrary Loan Department at Miller Library, Western New Mexico University, Silver City, New Mexico. Without their services, this book would have been impossible to write.

ONE

The Physics of Beauty

My GRANDMOTHER in Kansas had a large garden, which she used to provide flowers for my father's grave. We would cut bouquets of snapdragons, zinnias, and cosmos and put them in a coffee can set in the ground near the headstone. My father died when he was thirty-two years old. Where I live in Silver City, New Mexico, parents decorate the graves of children with holiday ornaments: Easter eggs, Christmas trees, a plastic wreath, a Valentine heart. Some parents do this for years and years after a child has died.

My grandmother put flowers on graves until she died, at the age of ninety-one: great glowing marigolds for her youngest boy, Milburn Grant Apt, weighty white chrysanthemums for her husband, Oley Samuel Apt.

Why do we give flowers to the dead? Why do we give flowers to the grieving, the sick, the people we love?

Fifty thousand years ago, the Neanderthals, too, buried their relatives with hyacinth and knapweed.

What are we offering?

Flowers are not symbols of power. Flowers are too brief, too frail, to elicit much hope of eternity. In truth, flowers are far removed from the human condition and from all human hope.

For a moment, in that moment, flowers are simply beautiful.

IN HER ESSAY "Teaching a Stone to Talk," the writer Annie Dillard complained, "Nature's silence is its one remark, and every flake of world is a chip off that old mute and immutable block. The Chinese say that we live in the world of ten thousand things. Each of the ten thousand things cries out to us precisely nothing."

Annie Dillard believes we silenced the world when we agreed that the world wasn't sacred. Most of us recognize this loss. The trees aren't speaking to us anymore.

My own experience has been somewhat different. Nature has never been silent for me. Nature whispers in my ear all the time, and it is the same thing over and over. It is not "Love." It is not "Worship." It is not "Psst! Dig here!"

Nature whispers, and sometimes shouts, "Beauty, beauty, beauty, beauty."

I am walking up a steep slope in the Sonoran Desert through sweeps of wildflowers. Someone is talking to me about pollination biology. I cannot listen as we walk uphill, because I am being knocked out by the flowers. I am breathing hard because of these flowers. I am an overexcited puppy. My tail knocks over the furniture.

This is classic arboreal desert: massive saguaro, numinous cholla, virile barrel cactus. Each of these sits apart from the other plants, showing off, in perfect composition. Red penstemon, yellow daisies, orange poppies, purple flax explode in the gravel, rippling like banners up hills and down arroyos. Their colors are the visible metaphor of joy. Flowers are celebratory. I have been invited to a party.

I feel a painful nostalgia. I used to live here, in this homeland, in this desert, in these hills, among the flowers. If I lived here still, I would be happy. I think to myself, "What went wrong?"

When Nature whispers beauty, I do not always respond well. Feverishly I want to get inside. I bang at the glass. It is *so* beautiful. It is too beautiful.

Only rarely do I feel calm, equal to the occasion. Then I am, myself, transparent.

I STOP IN A NEIGHBOR'S YARD to admire a sunflower. Its petals form a mandala, a wholeness made of many parts, just as the sunflower is made of many small flowers. In the center, each tiny "disk flower" has fused anthers that produce pollen, a female stigma that receives pollen, and a female ovary containing the ovule that will become a seed. If all goes right, each disk flower will pass on its pollen to a bee or other insect. Pollen is a food wonderfully nutritious and invariably messy. No matter how hard the creature tries, pollen grains stick to its legs, thorax, head, back, or under the wings. Eventually, some grains containing the male sex cells will dislodge on the female stigma of another disk flower. In a perfect world, each disk flower will be fertilized with pollen brought to it from another disk flower. Each ovule will form a sunflower seed.

Meanwhile, along the center's edge, the "ray flowers" unfurl one by one a single petal that with other ray flowers forms the larger circle. This is the ring of light that attracts the bee. A sunflower, like a daisy or dandelion, is really an inflorescence, a group of individual flowers acting together as a community.

These petals are an uncompromising yellow-orange. The color seems to contain all the energy this planet will ever need. This color could power a nuclear reactor. It rings like a carillon. It hits me, with a little punch, in the solar plexus.

The smell of the sunflower is more subtle. Bending closer, I breathe in earth and leaves and a delicate tang. There are odor molecules I recognize but cannot easily name: terpene, camphene, limonene. There are odor molecules I do not recognize and can barely smell. There are odor molecules I will never know because I cannot smell them.

I know that the sunflower is beautiful. I know this chemically. I know beauty, even though I do not know

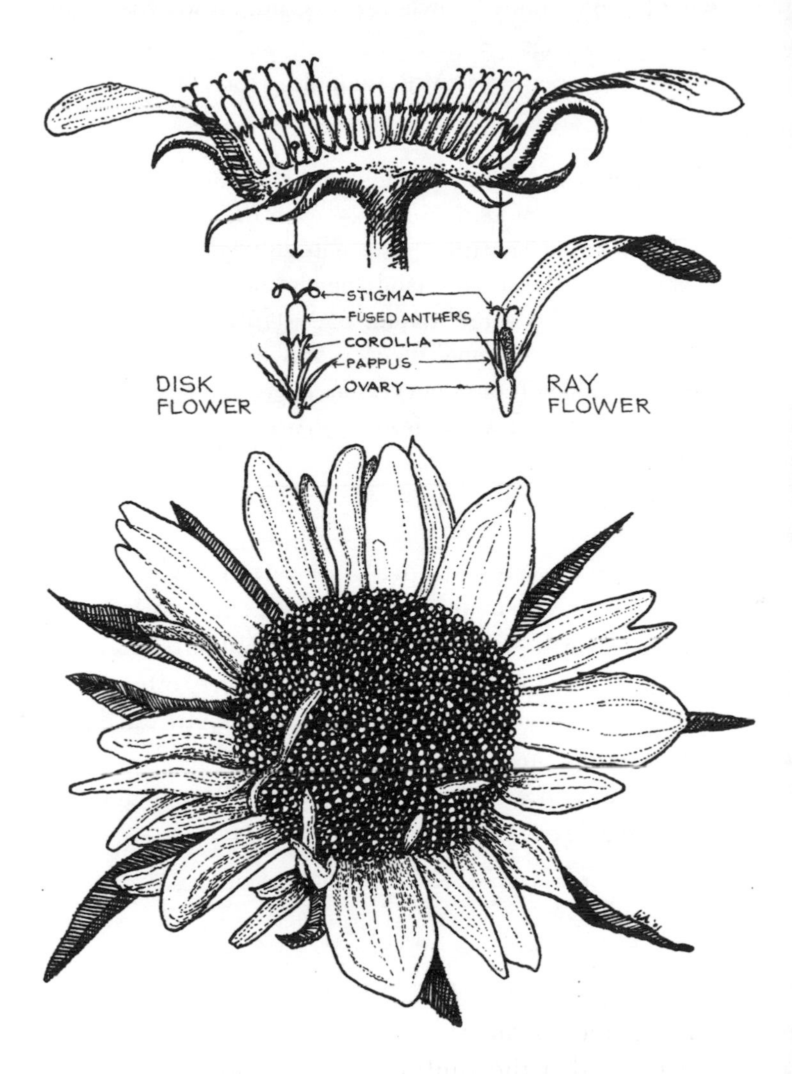
STIGMA
FUSED ANTHERS
COROLLA
PAPPUS
OVARY
DISK FLOWER
RAY FLOWER

Helianthus

what to do with my knowledge. I do not know what to do with my feelings.

The conservationist Aldo Leopold wrote:

> The physics of beauty is one department of natural science still in the Dark Ages. Not even the manipulators of bent space have tried to solve its equations. Everyone knows that the autumn landscape in the north woods is the land, plus a red maple, plus a ruffed grouse. In terms of conventional physics, the grouse represents only a millionth of either the mass or the energy of an acre. Yet subtract the grouse and the whole thing is dead.

Subtract flowers from the world and the whole world is dead from a human point of view. The nonflowering plants on earth include the mosses, liverworts, conifers, cycads, ferns, and gingko trees. Almost every other plant, everything we and other animals eat, requires a flower for reproduction.

We know that flowers are beautiful. We forget that they are also essential.

WE ARE BEGINNING TO EXPLORE the physics of beauty. Philosophers and scientists have come together to name certain universal themes.

The universe tends toward complexity.
The universe is a web of relationship.
The universe tends toward symmetry.
The universe is rhythmic.
The universe tends toward self-organizing systems.
The universe depends on feedback and response.
Thus, the universe is "free" and unpredictable.

The themes of the universe may be the elements of beauty. Certainly, they are the elements of flowers.

FLOWERING PLANTS HAVE RADIATED around the world to become our most diverse and complex form of vegetation. Flowers dominate every landscape except coniferous forests and lichen-based tundra. They astound us with their variation. We crush the tiniest ones under our feet and hardly notice the spiky blossoms of grass. Instead, we admire the giant arum (*Amorphophallus titanum*) that grows three feet across, its lip four feet from the ground, its central spire nine feet tall.

Early explorers thought the arum was pollinated by elephants who came to drink water collected at its base. Absently, the great creatures rubbed their heads against the pillar of pollen.

Elephant pollination is botanical kitsch. But flowers are pollinated by all kinds of animals, by marsupial mice and miniature thrips, by birds and lizards and butterflies and gnats and cockroaches and squirrels. A flower in Africa is pollinated by giraffes. The giant arum is pollinated by carrion beetles.

Like the arum, most flowers are one half of a partnership. They depend on a species extraordinarily different from themselves, someone who will carry their male sperm to another flower and bring compatible sperm to the egg in their ovary.

Some flowers depend on the wind. Flight is their means to reproduction. Is the Greek story of the North Wind, who could fertilize mares, any stranger? Are tales of Spider Woman or Moses parting the Red Sea any stranger?

The physics of beauty requires math. The sunflower has spirals of 21, 34, 55, 89, and—in very large sunflowers—144 seeds. Each number is the sum of the two preceding numbers. This pattern seems to be everywhere: in pine needles and mollusk shells, in parrot beaks and spiral galaxies. After the fourteenth number, every number divided by the next highest number results in a sum that is the length-to-width ratio of what we call the golden mean, the basis for the Egyptian pyramids and the Greek Parthenon, for much of our art and even our music. In our own spiral-shaped inner ear's cochlea, musical notes vibrate at a similar ratio.

The patterns of beauty repeat themselves, over and over.

Yet the physics of beauty is enhanced by a self, a unique, self-organizing system. Scientists now know that a single flower is more responsive, more individual, than they had ever dreamed. Plants react to the world. Plants have ways of seeing, touching, tasting, smelling, and hearing.

Rooted in soil, a flower is always on the move. Sunflowers are famous for turning toward the sun, east in the morning, west in the afternoon. Light-sensitive cells in the stem "see" sunlight, and the stem's growth orients the flower. Certain cells in a plant see the red end of the spectrum. Other cells see blue and green. Plants even see wavelengths we cannot see, such as ultraviolet.

Most plants respond to touch. The Venus's-flytrap snaps shut. Stroking the tendril of a climbing pea will cause it to coil. Brushed by the wind, a seedling will thicken and shorten its growth. Touching a plant in various ways, at various times, can cause it to close its leaf pores, delay flower production, increase metabolism, or produce more chlorophyll.

Plants are touchy-feely.

They taste the world around them. Sunflowers use their roots to "taste" the surrounding soil as they search for nutrients. The roots of a sunflower can reach down eight feet, nibbling, evaluating, growing toward the best sources of food. The leaves of some plants can taste a caterpillar's saliva. They "sniff" the compounds sent out by nearby damaged plants. Research suggests that some seeds taste or smell smoke, which triggers germination.

The right sound wave may also trigger germination. Sunflowers, like pea plants, seem to increase their growth when they hear sounds similar to but louder than the human speaking voice.

In other ways, flowers and pollinators find each other through sound. A tropical vine, pollinated by bats, uses a concave petal to reflect the bat's sonar signal. The bat calls to the flower. The flower responds.

THE MORE WE LEARN about flowers, the less silent they are. Perhaps all this listening is a way for the trees to speak to us again.

I can still smell my grandmother's garden.

We are just beginning to understand what we have always loved.

TWO

The Blind Voyeur

WE WALK THROUGH A FIELD of wildflowers. Sweeps of purple run up a hill. Stopping, looking closer, we see red skyrocket, orange globemallow, blue flax, yellow marigold. Flowers surround us with brilliant pointillism. Something in our chest lightens, dislodges, and rises into song. We want to sing like a bird.

We love flowers, obviously, because we love color. Human eyes process reflected light and pass that information on to the brain, where the real perception of color takes place. Color becomes wrapped in emotion and thought. Yellow is cheerful. Gray is sad. White is spiritual. People who have lost color sight describe a tear in the fabric of the world. One man saw his wife and friends as "animated gray statues." Food and sex became disgusting. Life seemed utterly wrong, dirty, unnatural.

Most of us, most of the time, take color for granted. We hardly notice the gorgeous, cerulean sky. We are habituated to green, the miracle that sustains us. It takes a hot-pink geranium to break through our calm. We exclaim at the velvet red of a rose. We delight in the sudden orange, the streak of blue.

Over 250,000 species of plants produce flowers, a vast array of color, scent, and form. This is pure spectacle, worthy of P. T. Barnum's greatest show on earth.

But human beings are not the intended audience. We sit in the theater, applauding and adoring, yet we don't understand most of the performance. We miss some of the best tricks. Flowers have patterns we cannot see, and they reflect colors we cannot imagine. The red poppy is not red to the bumblebee. The yellow cinquefoil may not be yellow to the butterfly. The purple snapdragon shimmers oddly.

Surrounded by flowers, witness to glory, we feel inspired. We feel grateful.

Unknowing, we are not much more than blind voyeurs.

I LIE DOWN IN THE GREEN grass, cuddling up to a patch of daisies, their centers the color of egg yolk, their petals a soft, milky white. Nearby, a red skyrocket (*Ipomopsis aggregata*) flaunts its long, trumpetlike, fused petals that end in a five-pointed, star-shaped opening. I only have a few minutes. Ants will begin to crawl up my ankle. Spiky leaves will tickle my skin. I will feel a growing uneasiness, so close to the ground, less than a foot high. I will

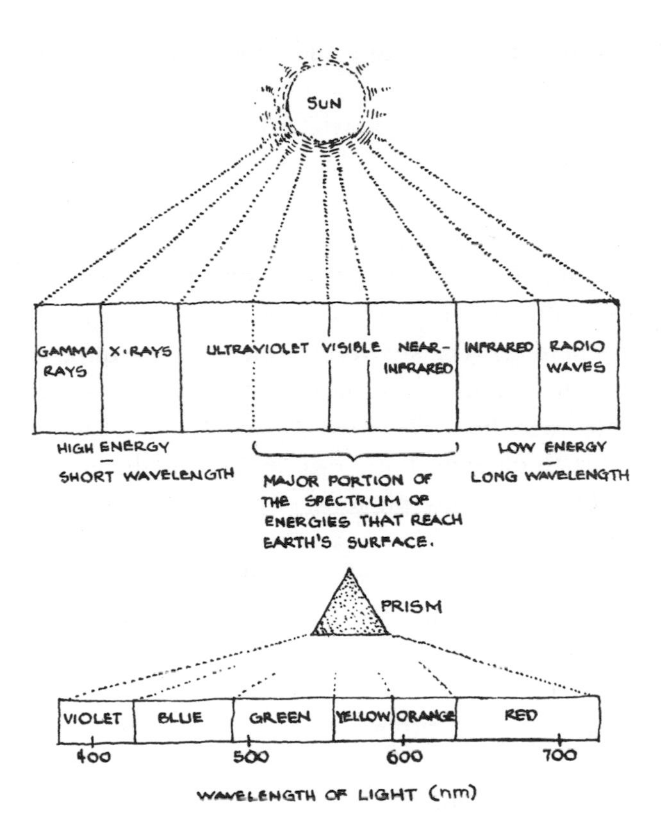
SUN
GAMMA RAYS
X-RAYS
ULTRAVIOLET
VISIBLE
NEAR-INFRARED
INFRARED
RADIO WAVES
HIGH ENERGY
SHORT WAVELENGTH
MAJOR PORTION OF THE SPECTRUM OF ENERGIES THAT REACH EARTH'S SURFACE.
LOW ENERGY
LONG WAVELENGTH
PRISM
VIOLET
BLUE
GREEN
YELLOW
ORANGE
RED
400
500
600
700
WAVELENGTH OF LIGHT (nm)

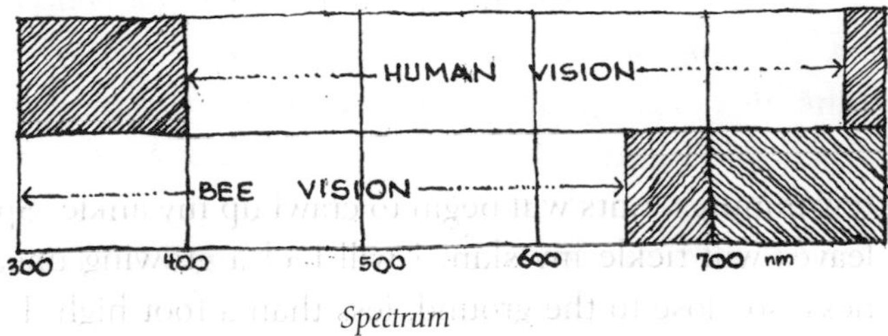
HUMAN VISION
BEE VISION
300
400
500
600
700 nm

Spectrum

feel, soon, the desire to stand and reclaim my bipedal perspective.

For a few minutes, the white petals of the daisy hold me. The smell of earth and leaves is familiar. I am being rocked, lulled, slipping into a dream as the sun drenches this meadow with radiant energy that moves toward me in rhythmic waves. The longest wavelengths are radio waves, infrared, and near-infrared. The last is the heat on my bare leg. The shortest wavelengths are ultraviolet, X rays, and gamma rays; most of these will never reach the earth's surface.

Between ultraviolet and near-infrared is the spectrum of visible light, those photons, or packets of energy, with wavelengths that the human eye can see. We perceive these different wavelengths as different colors. At one end of the spectrum is violet. At the other end is red.

When I move slightly, the petals of the red skyrocket loom large, filling my vision. In the cells of these petals are pigments that either absorb or reflect the different wavelengths of light. The pigments of skyrocket reflect back photons in the red range. They absorb most of the other wavelengths, and I do not see these colors.

The flower is keeping these colors.

What I see is the reflected red light that enters my eye, where the pigments there transform it into electrochemical energy, which is sent to my brain. I think, "Scarlet." I think, "Matador."

Although I understand all the words that explain light and the act of seeing red, the event itself is so fast and so complex that I do not understand, really, the sum of my words.

I move my head again, toward the white daisy.

In some white flowers, pigments in the cells of the petals reflect back all of the visible spectrum: red, orange, yellow, green, blue, and violet. When all the colors are reflected back from an object, we see white.

Most white flowers do not rely on pigment. Instead, they have petals filled with air spaces that reflect light. For the same reason, snow looks white because of air spaces in the frozen crystals. Different arrangements of floral cells can cause a diffuse reflection or a high refraction, a velvety matte look or a shiny brightness. If you squeeze the petal of a flower with air spaces, the air will be expelled and the limp petal will look colorless.

When the entire spectrum of visible light is absorbed by a petal, or by anything else, we see the color black. We do not see many black flowers. In 1939, one was reported to be growing in northern Oaxaca, Mexico. Fifty years later, a botanist went in search of the flower (*Lisanthius nigrescens*). He wrote that its buds were "glistening drops of coal oil," the open blossoms, an inch long, "bells of black satin." In the laboratory, he found that the petals were producing massive amounts of pigment, which absorbed wavelengths from red to violet at an incredible rate. No one knows what pollinates this flower. No one knows why a flower would dress in black.

Green, of course, is a color I see everywhere in this meadow—the dark green of the daisy's leaf, the light green of the skyrocket's stem, the emerald green of new grass, the distant green of juniper trees and ponderosa pine. In school, most of us learned about photosynthesis, a subject we might have done better to study in church. The pigment chlorophyll converts light to energy. We all depend on this act of grace.

Chlorophyll absorbs best at wavelengths of violet-blue and orange-red. It reflects back and does not use wavelengths of green. Biologists think they know why. The ancestors of plants evolved at the bottom of a sea, under swarms of aquatic bacteria that were already absorbing and using the green wavelength. Plantlike cells that developed pigments that could absorb and use the remaining spectrum survived (and hence reproduced) better than other cells. Once on land, in full sunlight, plants didn't need to become more efficient. Plants continued to reflect green rather than absorb all wavelengths. Perhaps for this reason, we do not walk under trees of black satin or picnic in fields of coal-dark grass. I am grateful for serendipity.

A bee comes to visit the daisy. The insect lands, plopping, thumping, shaking the flower. The daisy seems suddenly alert. The daisy seems suddenly relieved.

Rocked in rhythmic waves, lulled by the sun, I would be the daisy's lover. I would embrace the red skyrocket, the purple verbena, the orange firewheel, the blue flax, the yellow groundsel. I yearn for the daisy, for color, for love. But I have no real sense of commitment. I have no intention of pollinating these flowers. And these flowers have not been waiting for me.

THE BEE IS a serious pollinator. Bees include over 25,000 species, big, small, stingless, aggressive, social, and solitary. We have been studying bees for a long time, especially honeybees, and we are constantly amazed at what they can do. These tiny things dance. They communicate. They remember. They learn. Bees have been called the intellectuals of the insect world. (Butterflies, unfairly, are known as the dumb blondes.) Bees have taught us never to underestimate.

Honeybees have three types of photoreceptors, or light-sensitive cells, with peak sensitivity in the areas of ultraviolet, blue, and green. Human beings see best in the areas of blue, green, and red. The reflection or absorption of ultraviolet light constantly affects how a bee sees the world.

In my mountain meadow, I am struck by the popularity of yellow. In a range of species, in every shape and size, yellow flowers are everywhere. Yellow is bright. Yellow is cheerful. Was it on sale?

To me, many yellow flowers simply look yellow, like the flowers of wormseed mustard, rape, and field mustard. But since each of these flowers reflects ultraviolet light differently, a bee will see three different colors.

In human sight, the two ends of the spectrum, violet and red, combine to form purple. In bee sight, the two ends of the spectrum, ultraviolet and orange-red, combine to form what some scientists call "bee-purple." They could as reasonably say "foog" or "orumpho," a more alien word.

When all the colors of the human spectrum are reflected back from an object, we see white. When all the colors of a bee's spectrum, including ultraviolet, are reflected back from an object, the bee sees "bee-white," a color we might not recognize.

To a bee, most human-white flowers look blue-green, while the green leaves of this daisy probably look gray. Although bees do not see far in the red range, only a few flowers absorb a bee's entire spectrum and look "bee-black." Red flowers that reflect some blue light look blue. Red flowers that reflect ultraviolet look ultraviolet.

What color is ultraviolet? What color is blue and ultraviolet? What color is yellow and ultraviolet? What colors, really, are all these flowers in this meadow?

We don't know, since we can't see them.

Perhaps nothing strains our imagination so much as an experience outside our evolution. We don't have the chemistry. We don't have the neurons. We can't make the color happen in our brain.

The skyrocket bobbles in a breeze. This close, the vivid red shows tints of orange. Small, white spots decorate the inside of each flower's starlike opening and run deep into the corolla, the trumpet-shaped fused petals. Looking deeper still, a pink glow is visible.

Once a flower has gotten a pollinator's attention, it may begin to use color differently. Guide marks, like these small, white spots, direct an animal to the source of nectar or pollen. Rings at the flower's center are the bull's-eye. Lines and arrows point dramatically. The yellow streak on an iris is a landing strip, signaling small aircraft. Rows of green dots lead the way in a marsh gentian. Orange markings do the same in a monkeyflower.

Read the sign, please. Don't wait to be seated. Dinner is this way.

Different parts of a flower, differently colored, also reflect or absorb ultraviolet light. Some guide marks are completely invisible to the human eye.

I feel the frisson of a parallel world: flowers glowing in strange colors, flowers marked by strange patterns. Briefly, I want to see what the bee sees.

Let me slip under the surface of this dream.

Let me lift this veil before my eyes.

THE FOSSILS OF THE EARLIEST KNOWN FLOWERS are about 120 million years old. Bees have been around much longer than that, and bees probably had color vision long before the appearance of flowers. In this evolutionary dance, the flower first courted the bee. The color of the flower is part of the invitation. Here, here, here, the flower hums. Come to *me*.

Of course, flowers evolved to attract a variety of insects. With a visual range of ultraviolet to brilliant red, a butterfly sees color better than a bee, and better than you. Some moths see just as well as butterflies. Beetles are important pollinators, and the dung beetle can distinguish yellow, orange, and violet from blue, as well as yellow-green and light green. Most or all flies see in color. Tiny thrips, which feed on pollen, respond best to blue-green, blue, and yellow. Other insect pollinators include wasps, earwigs, cockroaches, book lice, grasshoppers, crickets, and lacewings. Researchers have yet to investigate their color sight.

Many flowers are pollinated by birds, which have wonderful vision. Male and female starlings, with their black, iridescent plumage, look alike to us but very different to each other. The starlings are attracted to ultraviolet patterns not in our bird books. Like butterflies, birds can easily see red. In the Americas, red flowers are commonly visited by hummingbirds; in areas like central Europe, which do not have pollinating birds, red flowers are more rare.

Mammals also pollinate plants. Nocturnal bats usually drink nectar from white or cream-colored flowers that stand out in the darkness. Many shrews, small marsupials, and rodents feed in twilight and prefer light colors. Nectar-rich flowers that attract mammals mainly through smell tend to be dingy or dull-colored, growing low to the ground.

These kinds of patterns—white for bats, red for birds—are known as pollination syndromes. For a while, scientists thought that floral cues of color, scent, and form were a kind of crossword puzzle: all the cues formed a pollinator that was born hardwired to respond to a yellow marigold or a red skyrocket. The blue, sweet-smelling flower with a deep, narrow tube was cross-matched to a butterfly. A red, unscented, trumpet-shaped flower was pollinated by hummingbirds. A greenish, skunky blossom attracted flies.

Most scientists today play down the role of pollination syndromes and innate preference. Birds and bees and butterflies are just too flexible. They are selfish and pollinate the flower they like best, or the flower they can find, not the flower they were born to pollinate. They become field experienced. They respond to choice and chance.

In one experiment, young swallowtail butterflies were presented with paper flowers in many different colors. They preferred yellow. Blue and purple came in second.

Next, the researcher manipulated a certain wildflower, which has both yellow and magenta blossoms. She presented the butterflies with real yellow flowers without nectar and real magenta flowers with nectar. (To drain nectar from a flower, the researcher brought in advance teams of hungry butterflies. To make sure the flower was empty, she inserted tiny paper wicks into each tiny nectar tube.)

Yellow remained a first choice. But after only ten flower visits, most butterflies switched to magenta. Now, these experienced butterflies were given a third choice: yellow flowers with nectar and magenta flowers without. They adapted again, quickly.

The same thing happened with hummingbirds. Hummingbirds like red. But if I were to paint half the skyrockets in this meadow white, and remove nectar from the remaining skyrockets, the hummingbirds would decamp to white.

Color as an invitation is too subtle. Color is an advertisement. Red is a billboard sign.

Coke! Pepsi! Eat here!

The product now has to live up to the hype.

A few flowers rely on false advertisement. Their color or scent promises a reward that is never given. These flowers do depend on hard-wiring, the innate preference of newly hatched customers. Some good mimics can even fool a pollinator over and over.

Flowers bobble. Flowers glow. Flowers shout.

Come to *me*. Come to *me*. Come to *me*.

SCIENTISTS WHO STUDY FLOWERS actually find themselves painting the petals of a red skyrocket white. They use an acrylic that they claim does not hurt the flower or affect the pollinator. Then they stand back and watch. What will visit the flower now?

Flowers have also tried changing colors, in their own experiments, with good results. A flower may change color as soon as it is fertilized, or it may change automatically with age, when it is most likely to have been fertilized. The new color tells a pollinator that its services are no longer required. The bee can go elsewhere, preferably to a flower on the same plant or inflorescence.

A more obvious strategy might be to have the flower drop off and die. But if reproductive changes are still taking place, parts of the flower may still be required. A post-change flower can also be useful to a plant that has flowers not yet fertilized. The large floral display continues to attract pollinators from a distance.

Color change is surprisingly common and surprisingly unpredictable. Within a family, it can occur in some genera and not in others. Within a genus, it can occur in some species and not in others. Within a species, it can occur in some individuals and not in others.

The mechanism of change varies. A young flower (*Bauhinia monandra*) from the West Indies is white, with a large, red spot in the middle of the central petal. As the flower ages, the central petal curls back and covers the red spot. Meanwhile, the four side petals turn pale pink. Now the entire flower looks pink. It is a strong signal: I'm old. Don't touch my stigma.

In the same genus, the appearance of a pigment causes a yellow flower to turn red. The disappearance of a pigment causes a white flower to lose its ring of yellow.

Changes in a flower's pH can also affect color, turning pink flowers blue and blue flowers pink.

Flowers pollinated by night-flying moths or bats often change from white and cream to dull red, gold, or purple. Receding into the darkness, the post-change flower may still produce scent, helping attract visitors to other flowers on the plant.

The banner petal on a white lupine is now purple.

A field of white lilies is pink and red the next morning.

A yellow blossom is no longer yellow.

Messages are being sent, information exchanged. The code is in color. The colors are fleeting.

We walk through a field of wildflowers and we love the yellow cinquefoil even though it is really foog and we love the poppy even though it is ultraviolet. We are blind voyeurs. We have been invited to a party, and it doesn't seem to matter that we fail to recognize the host or many of the guests, that we stumble about awkwardly, not knowing what we are not seeing. Our hearts are gladdened. We know what we feel. Flowers make us happy.

THREE

Smelling Like a Rose

IN THE AISLES OF EVERY DEPARTMENT STORE are products that use the brand names of rose, orchid, violet, honeysuckle, magnolia, narcissus, orange blossom, carnation, and hyacinth. We use these scents in our soaps and perfumes, bubble baths, lotions, shampoos, deodorants, and even air fresheners and cleaning products.

We want to smell like flowers.

We are no different from most cultures, ancient and modern. The Hindus and Egyptians worshiped their gods with fragrance. The Greeks specialized in perfumery. The Bible reeks of incense. Europeans thought eau de cologne could ward off the plague. Aztec noblemen carried fresh flower bouquets. In all of human history, there is hardly a time or place that does not reflect our preoccupation with smelling good.

Most perfume today has three odor groups, or notes. The top note comes first, with a floral highlight such as lilac or lily. The middle note provides body and uses the essential oils in jasmine, lavender, or geranium. The third, or base note, includes animal products, such as musk from rutting deer or the pasty fluid from the anal glands of civet cats. These last products also add the intangible qualities of body and warmth.

The human body has its own array of scent from glands scattered on the face, scalp, breasts, under the arms, and in the genital area. Oddly, humans are desensitized to this last smell. Long ago we suppressed our ability to sniff out ovulation. One theory suggests that when we began living in complex social groups, smells of sexual readiness threatened the pair bonding needed to raise children. Frankly, culturally, we are a bit disgusted by our own odor. We do not want to smell too human.

But we do want to smell like *something*. And we want to attract mates. So the top notes of perfume are from flowers that use scent to attract pollinators. The middle notes are oils and resins that resemble sex steroids. The base notes, at low concentrations, are obvious.

We don't want to overstate our case. We don't want to smell too much like a deer or civet cat.

We want to smell like a rose. We want to smell like orange blossoms. We want to smell like jasmine.

On their part, most flowers want to smell like food. Some flowers want to smell like a rotting corpse. Some flowers want to smell like excrement. Some flowers want to smell like fungi.

Flowers have their own agenda.

A SINGLE FLOWER CAN produce as many as a hundred chemical compounds, with smells mixing and combining in patterns that change over time, with parts of a flower smelling differently from other parts, with smells sending out a variety of signals: Lay your eggs here, nectar over there, eat now.

In large quantities, the chemicals that produce scent are often toxic. To protect the plant, they are stored as volatiles (oils that convert easily from a liquid to a vapor) in special cells, usually in the flower itself. The petal tissue might manufacture some of these volatiles; the reproductive organs may be responsible for others. The vegetative tissue of the plant also adds to a flower's fragrance, which is usually a blend of many odors.

Odor molecules are released through the process of evaporation. Once in the air, the molecules begin to move about randomly, farther apart from each other, until they are carried away from their source by the wind. For a while, however, there is a coherent trail of molecules, known as an odor plume. This plume has a destination. More often than not, it is meant to intersect an insect's antenna, which has hundreds of cells designed to catch it. The outline area of the antenna is the insect's nostril. In some moths this can be as large as a small dog's. A dog sniffs to breathe in smell. An insect waves its antenna.

Entering a plume, insects tend to zigzag, casting down or to the side when the odor is lost. When a zigzagging insect gets close enough to actually see a flower, it may suddenly "beeline" to its destination.

Flowers smell so good because insects smell so well. Some moths can detect a scent a mile away. Most moths, like poodles, smell just about everything. Other pollinators, especially bees, also distinguish and remember odor. In response, flowers may have evolved a complex mix of smells, and this may encourage what botanists call flower constancy.

Flower constancy is the "loyalty" of a pollinator to a specific flower or species. First, a flower species "wants" to smell and look different from competing species. Second, a flower wants to attract a pollinator that will recognize and remember the difference. Finally, a flower wants that pollinator to be loyal, flying off with a load of pollen to fertilize a compatible flower.

For their own reasons, insects seem to oblige. Even when other flowers are blooming, a bee may continue to visit the familiar red clover or pink four-o'clock. The flower gets fertilized by a similar flower, and the bee becomes adept at handling that species. Because a honeybee might visit five hundred flowers in one foraging trip, any savings in energy or time adds up quickly. We choose the same strategy when we shop, every day, at the same grocery store or drive the same route to work.

Because different flowers open and release their scent at different times, an insect can have many loyalties. The blue chicory has nectar in the morning. Red clover is best after lunch. The four-o'clock opens in late afternoon. The evening primrose follows the four-o'clock.

A honeybee's memory for odor is linked to a certain time of day. Typically, bees seem to "trapline" a series of flowers, moving to each appropriate one at the appropriate hour, and then heading straight for home.

The timing of scent production is part of a flower's reproductive success. Some flowers, like roses and clover, are scented only in the day. Some flowers are scented only at night.

There are smells that you and I will never know, because we are not nocturnal. There are smells like maps to the country of loyalty.

EVERY YEAR, worldwide agriculture produces over 100 million tons of cane and beet sugar. The Australians, Irish, and Danish eat one hundred pounds of refined sugar per person per year. Americans eat a little less. Nectar is mainly sugar water, sometimes containing sucrose or cane sugar, sometimes a mix of sucrose, fructose, and glucose. Most of us understand the butterfly. We would also extend our mouthpart, automatically, at the equivalent of a candy bar.

In concealed nectaries or in open pockets, nectar is the pollinator's reward. In different species, nectar can be secreted by almost any part of the flower. Nectar-rich flowers often have a strong, sweet fragrance that does not necessarily come from the nectar. (Bird-pollinated flowers also provide nectar and are relatively unscented because birds do not have a good sense of smell.) The flower's strong perfume is first a lure, a dinner bell, and an advertisement.

Up close, scent marks on the flower, like visual guide marks, may further direct an insect to the source of food. The subtle smell of nectar may help some bees determine if the flower is full or empty.

Flower mites also use nectar to recognize their host species. Mites ride from flower to flower in the nostrils of the hummingbird; when they smell the right nectar, they gallop home down the bird's beak.

In some flowers, pollen is the reward, and the odor of pollen is the primary scent. This is particularly true in plants visited by pollen-eating beetles. Bees, too, are good at smelling and distinguishing different pollen from different flowers.

Perhaps the best image for the smell of pollen is the kind of breakfast only a farm worker should have: eggs, bacon, ham, cheese, potatoes, biscuits, and gravy. The steaming plate sends out a strong odor plume.

Pollen can also be sexy. In the sunflower moth, when virgin females are exposed to the odor of pollen, they begin signaling for males earlier and spend more time signaling. Later, more of their eggs mature.

This interplay of food, scent, and sex is a common theme. Some flowers smell like the sex pheromones of a butterfly; the male carpenter bee attracts females with a pheromone that smells like a flower—good enough to eat. (A chemical is considered a pheromone when it is used as communication between at least two members of a species.) Through millennia of mimicry and exploitation, flower volatiles and insect pheromones have co-evolved, with flowers imitating pheromones and pheromones imitating flowers.

We want to smell like a rose. We want to smell like a butterfly. We want to smell like an insect pheromone.

It doesn't stop there. A major ingredient in the sex pheromone of many moths is also found in the sex pheromone secreted in urine by female Indian elephants. The urine is meant to attract the attention of a bull—the bigger the better.

In one experiment, women who sniffed musk, the sex attractant of Himalayan deer, developed a shorter menstrual cycle, ovulated more often, and conceived more easily. The smell of musk resembles the smell of steroids found in human urine. The chemical structure of steroids like testosterone resembles plant resins like myrrh. We use these resins in our perfumes, just as we use the volatiles of flowers.

The fact that so many things smell like each other is partially explained by nature's efficiency. A compound that works here will also work there. We all came from the same primordial soup. Poets, who equate one thing with another, often echo scientists. Similes are real. Metaphors are chemical.

In the Bible's lyric poem the Song of Solomon, odor is the language of love: "My beloved is unto me as a bag of myrrh that lieth between my breasts. My beloved is unto me as a cluster of henna flowers in the vineyard of Engedi."

The flowers of henna, lime, and chestnut smell like semen. Myrrh has a smell compared to the oils secreted by glands in the human scalp.

WE WANT TO SMELL LIKE A ROSE. We want to smell like a henna flower.

But we don't want to smell like the largest inflorescence in the world, the nine-foot-tall giant arum, mythically pollinated by elephants, which gives off a stench so revolting it has made men grow faint.

We don't want to smell like the dead horse arum, a flower that evolved near gull colonies and came to resemble the rotting carcass of a bird. This arum is round and plate-sized, grayish purple, blotched with pink, and covered with dark red hairs, or trichomes. Its smell of decay draws blowflies, which come to feed and lay their eggs. The flies crawl into what appears to be an empty eye socket or an inviting anus, deep into the flower, where they are trapped, the exit closed by bristly hairs.

Fed by nectar, the flies lay eggs that will later hatch and die for lack of food. Suddenly, the flower releases its pollen, which covers the blowflies. The bristly hairs wilt. The flies crawl out again.

Other flowers pollinated by flies and beetles can smell variously like dead animals, rotting fish, or dung. Colors of red, purple, and brown add to the effect. Dark spots or warty areas look like clusters of insects already feeding. The common names of a plant tell the story: skunk cabbage, corpse flower, stinking goosefoot.

We don't want to smell like a dead horse arum. But we do want to smell like jasmine, which also has a distinct fecal odor beneath the top notes of cloying sweetness. At very low levels, at levels that reach into childhood, at the level of the unconscious mind, at a level that defines our kinship with the rest of the world, urinary and fecal smells are commonly added to our best perfume.

MOST FLOWERS SMELL like a restaurant. They use scent to signal an insect that they have food or to deceive an insect into believing that they have food.

Some flowers smell like home, a good place to raise a family. A fungus gnat lays its eggs where the eggs can hatch and eat the fungus. Plants that mimic fungi grow low in the forest, with dark purple or brown flowers. Fleshy parts of the flower seem particularly to attract the fungus gnat. One orchid has a creamy, gill-like area that resembles, precisely, the underside of a mushroom.

A few flowers are drag queens that advertise for sex and use scent as part of their costume. A Mediterranean orchid has an oval, convex lip that glistens metallic violet-blue. Its narrow, yellow border is fringed with reddish hairs. Dark red, threadlike upper petals move in the wind like an insect's antenna. The orchid looks and smells like a female wasp. When the male wasp lights and tries to copulate, pollen is transferred onto its head.

Pseudocopulation is rare but not unique. Around the world, certain bees, wasps, and other insects are trying to mate with flowers that appear to be what they are not. (The bee fooled by a mere flower may be lucky. The larvae of blister beetles also lump together to look and perhaps smell like a female bee. These larvae first attach themselves to a male and then try to find their way into a young bee nursery, where they will feed on pollen.)

Occasionally, a cigar is just a cigar. The boldest and most bizarre use of smell may be "perfume flowers," whose scent tells a male euglossine bee that this flower has . . . scent. Like someone at a department store, shopping for a very important night, the bee mops up the perfumed liquid with feathery brushes on its front feet. The smell is stored in a pouch in a back leg and combined with other odors to create a private irresistible pheromone.

SMELL IS A COME-HITHER. It can also be a go-away. Some fertilized flowers change their scent to signal pollinators to pollinate elsewhere. Many flowers stop producing scent altogether, the most definitive no.

Pollinators use scent, too. Bees secrete a pheromone to mark recently visited flowers. The short-term odor is a memo to oneself: This flower is out of nectar. Other bees respond to the smell. No one wants to crawl up an empty corolla tube.

THE MOST EXPENSIVE PERFUME in the world, Joy, mixes a little jasmine with lots of rose. Roses have always generated passion. The Romans celebrated the holiday Rosalia to excess. When the prophet Muhammad ascended to heaven, drops of his sweat fell to earth and turned into roses. The early Christian rosary had 165 dried and rolled-up rose petals.

The smell of a rose is first absorbed by the mucous membranes in our nasal cavity. Next, receptor cells fire a message to the limbic system, an ancient part of our brain and the seat of emotion. Here, memories associated with smell decay more slowly than visual memories.

We want to smell like a rose. Of course we do. Everything else is smelling like something. Molecules everywhere are drifting into the wind, jostling against each other, grabbed by a sensory cell, by the antennae of a moth, by a dog's nose, by a lover's inhalation. We want to be part of that movement. We want to move. We want to be moved.

FOUR

The Shape of Things to Come

THE PASSIONFLOWER IN MY NEIGHBOR'S YARD looks constructed, designed by an engineer who has heard about flowers but never actually seen one, designed by a woman in love with helicopters.

The passionflower is layered. Five green sepals and five green petals form the base. A fringe of spiky modified petals swirls above like a sea anemone in concentric color: an outer ring of lavender, a ring of white, a wide ring of purple, a ring of green, a thin ring of purple, a ring of light green, a center of dark purple.

From this center rises a stalk almost an inch high. Five pads resembling the brake pads of a bicycle hang down. Their bottoms shine with yellow pollen meant to dust any bee or fly drawn to the mosaic of rings and the nectar at its base. Above the brake pads, three stigma lobes splay out like the rotors of a funny hat, a helicopter hat, a beanie hat.

The whole thing looks preposterous.

When explorers from Europe first saw a passionflower in the New World, they immediately sent one to the pope, claiming that the flower reminded them of the thorny wreath of Christ and his passion on the cross.

What on earth were they thinking? Perhaps, like me, they simply had to have some response, a metaphor to hold on to in the face of the passionflower's incomparable and willful existence.

It's a sea anemone. It's a helicopter. It's the passion of Christ.

THE PASSIONFLOWER IS a circle that can be divided into equal parts. Circle flowers are accessible to a variety of insects that simply land anywhere and walk to the flower's center. Circle flowers are easy to manipulate. They're democratic.

Botanists call the passionflower a *perfect flower* because it has both male and female organs. At the center of a perfect flower is the female organ, or carpel, which holds the ovules, or unfertilized eggs. The base of the carpel is the ovary. From the ovary rises a neck, or style, topped by one or more stigmas. The stigma receives the pollen grain. Sperm in the pollen grain will travel down the style to fertilize the eggs.

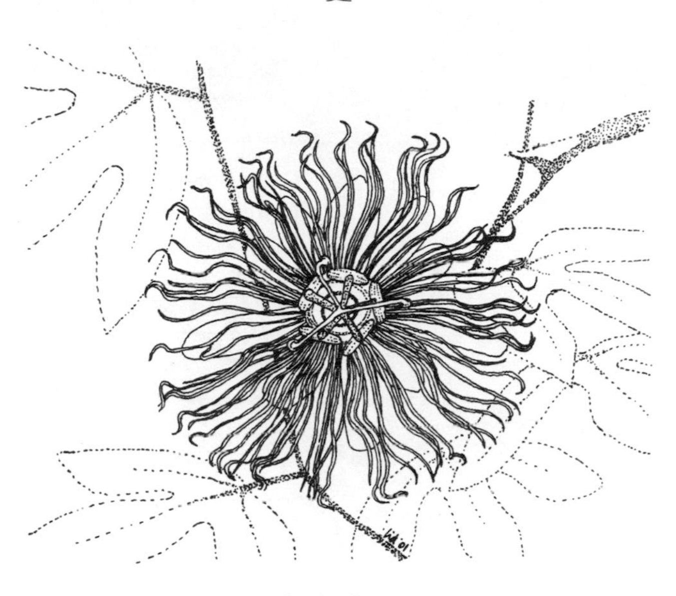

Passionflower

Surrounding the female carpel is usually a ring of male organs, or stamens. Each stamen is made up of a stalk ending in an anther, which produces pollen. The petals, collectively called the corolla, surround the stamens. The leaflike sepals that first enclosed the bud, collectively called the calyx, surround and grow under the petals.

If you look at flowers a lot, every day, like eating a good breakfast and getting regular exercise, you will remember these terms. Otherwise, you will not. You will remember the metaphor, a word like *helicopter*.

Some flowers do not have all these parts. They may have only the male or the female organ. They may have

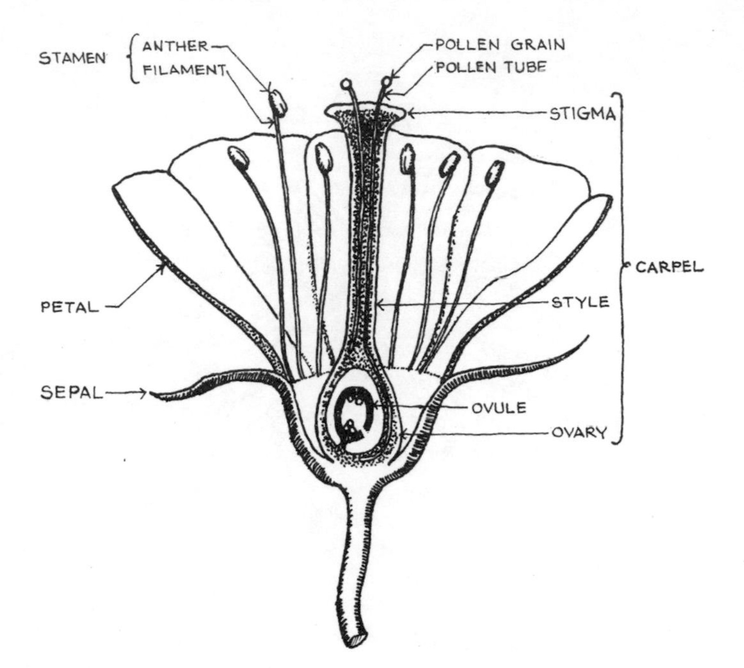

Parts of a Flower

only one carpel or many carpels. A carpel may have one ovule or, like some orchids, half a million.

Some circle flowers, such as daisies, are really inflorescences, or arrangements of flowers. At the center of these is a community of individuals. Each may have its own tiny carpel, stigma, stamens, corolla, and calyx.

Many flowers, of course, are not circles. Cut in half, a bilaterally symmetrical flower will have two mirror images. The lower part may look very different from the upper. In bilateral flowers, the petals have often fused to become a funnel, a bell, a trumpet, a pipe, a hollow slipper, a long spur, a mushroomy gill, or something that resembles a wasp, a bee, or something else. Stamens may be fused to the inside of the corolla or to other parts of the flower, like the ovary or style.

In all flowers, sepals can start acting like petals. When botanists cannot tell the difference, they call the sepal/petal a tepal or a petaloid sepal. (A tepal may be what really forms the base of the passionflower.)

In grouping plants, the assumption is that each species in a plant family evolved from the same ancestor. But within a plant family, flowers can include a startling range of flower shapes, from circles to gullets to spurs. Indeed, as flowers evolve through time, they seem to be in a state of constant transformation: melding, moving, merging, flowing.

This liquidity is entirely practical. Flowers change their shape in response to pollinators, predators, or the environment. A flower may "want" to attract a certain bee, defend itself against ants, or conserve water.

Bilaterally symmetrical flowers, for example, tend to control more precisely how a pollinator gets pollen and leaves it behind.

Many orchids have a beautifully decorated lower lip, which provides insects with a good place to land. From here, the animal stands and pushes its head or entire body into the flower's upper gullet. As the pollinator backs out, the pollen clings to the thorax, abdomen, or some other place likely to contact the stigma of the next orchid.

In a pea blossom, a single large banner petal signals the bee. Two smaller petals, or wings, surround a keel petal. When the bee lands on the keel, its weight causes the petal to dip. The enclosed stamens pop out and dust the bee.

In a delphinium, a long spur is attached to a miniature petal with wings. An outer ring of five tepals attracts the bumblebee and provides support as the insect inserts its head between the wings and flicks its long tongue into the spur, searching for nectar. Stamens at the top of the spur cover the insect's head with pollen. When hummingbirds pollinate the delphinium, they hover and make no use of the flower's support. Instead, the pollen rubs off on their beaks.

Form follows function. It's nicely alliterative. But flowers are a little more complex than that.

Flowering plants are called angiosperms, *sperma* for "seed" and *angeion* for "inside a vessel," because their closed, fleshy carpels protect the developing seed from predators and a hostile environment. Before angiosperms, gymnosperms (*gymno* for "naked") like conifers were the dominant vegetation. In the history of plant evolution, carpels proved to be a tremendous maternal leap. The angels in heaven cried out at this one. Hosannas resounded.

Obviously it is important how well a closed carpel protects its seed. In a few flowers, the ovary is high above the other organs and thus more vulnerable to attack. In roses and cinquefoils, the ovary is still high but is surrounded by other organs and petals that give it more protection. In orchids—and in the small disk flowers of a daisy—the ovary is covered by layers of fused tissues, buried deep and well defended.

Sometimes that defense is an example of form following function. But sometimes it is more accidental, a result of the fusion of flower parts for other reasons. As Peter Bernhardt points out in his book *The Rose's Kiss,* "An orchid flower needs a lot of fusion to hold a pollinator between its lip petal and column." Thus flower organs have united. In so doing, they covered the ovary. This is good. But it may have also resulted in other changes, other problems to solve, other advantages to exploit.

The flower keeps fiddling.

Form follows function.

It's mostly true.

I AM CUTTING INTO the tiny spur of a delphinium, pretending to search for that source of nectar that feeds the hummingbird. I use a little scalpel, a little pair of tweezers, and a magnifying glass. My fingers seem gigantic. I peer. I squint. I insert my scalpel. I will never see what I really want to see, what I fancy dwells at the bottom of the delphinium's spur.

What does evolution look like?

It is not so crazy to squint for it in a small place. You could well say that evolution is minuscule since it begins with changes in a gene or cell. As cells divide and duplicate themselves, the genes carried on their chromosomes must also divide and duplicate. In this process, the gene occasionally mutates, or changes. For the organism, the change may be good or bad or neutral. In any case, the duplicated gene is slightly different.

In cross-breeding, one set of genes from a parent combines with a set of genes from another parent to create a new individual. This results in even more change and more genetic variation within the group.

The process of natural selection takes over. A beneficial genetic change may help an individual survive and reproduce in a particular environment. The change may be passed on to the individual's offspring, who will have an advantage over other individuals. These changes accumulate, individual by individual, generation by generation, until the population itself has taken a new shape.

The idea bears repeating. Genetic changes that give the individual a closer fit to its environment and that therefore result in an individual's higher rate of survival and reproduction will gradually become the status quo.

A species can be broadly defined as a group of organisms able to breed with each other to produce a new, fertile generation. A species can change over time into another species. In this case, called phyletic evolution, only one species remains.

Speciation, however, is when a species splits into two species. Speciation is why we have roses and orchids and philodendrons. Natural selection is only a part of this process. Something else has to happen first.

Often, two or more populations become isolated from each other for a variety of reasons and in a variety of ways. A continent drifts north. An island rises from the sea. An asteroid crashes into earth. External forces separate populations, as might internal ones. The separated groups evolve along different paths. At some point, they become separate species.

In a study of monkeyflowers, researchers found that a change in one gene alone may have been enough to increase nectar flow and double hummingbird visits. Another small gene change altered the flower's pigment and reduced bee pollination by 80 percent. Relatively few genetic changes may be necessary for reproductive isolation—and for speciation.

Evolution can be "fast" or "slow," measured in hundreds or millions of years. It can begin with a gene or a volcanic eruption. The process has no direction, no purpose. Evolution requires the randomness of genetic change, followed by the decided nonrandomness of natural selection, complicated by the extreme randomness of external events.

Evolution is not hard to see, although it may be impossible to fully understand.

What does evolution look like? Look around. That tree. That bush. That insect. I can find evolution in the spur of a delphinium, just as I can find it everywhere in all the living world.

I will not find it in a dismembered flower.

And suddenly I feel like a serial killer, surrounded by body parts. I will see evolution only in the process of life, immanent, like some version of God. It is not the thing itself. It is not the tree. It is the shaper of trees.

It is the shaper of flowers.

Flowers pollinated by hummingbirds often have curved corolla tubes. This curve makes the bird's beak push against the tube and touch the anthers. In response, some hummingbirds evolved a matching curved beak, which made their feeding more efficient. In response, some flowers evolved even curvier corollas, which made the curved beak push again against the top of the corolla tube.

Some of us respond to this information with a kind of awe, as though we were hearing organ music, as though the light were filtering through colored window panes.

The flowers of the delphinium grow in a spiral on a tall spike. The flowers at the bottom of the spike are older, larger, and in a female stage. The anthers have shed their pollen and the mature stigmas are now receiving pollen. The upper flowers are younger, smaller, and in a male stage. The anthers are producing pollen and the stigmas are not yet mature.

The older, larger, lower flowers tend to have more nectar, and the bumblebee's strategy is to start at the bottom, forage up, and then fly down to the next delphinium. This works well for the bee in terms of flying costs and nectar reward; its energy use is most efficient. This also works well for the delphinium, for the bee is now carrying pollen to the female flower of another plant.

The delphinium has evolved to maximize its pollination, based on the bee's behavior. The bee has evolved to maximize its foraging, based on the plant's architecture. The match of delphinium and bee is not perfect, nor is it simple or unchanging. The delphinium also wants to attract other pollinators, and the bee wants to visit other food sources.

Concerning his own theory of evolution, Charles Darwin wrote: "There is grandeur in this view of life, with its several powers, having been originally breathed by the Creator into a few forms or into one; and that, whilst this planet has gone cycling on according to the fixed law of gravity, from so simple a beginning endless forms most beautiful and most wonderful have been and are being evolved."

Charles Darwin did not have too much trouble putting a creator into the heart of evolution. The pope, I presume, had no trouble seeing the passion of Christ in a passionflower. I have trouble with these things every day. I am a product of the last half of the twentieth century. Hoping to find God, I cut up the delphinium.

The shape of things now is not the shape of things to come.

FIVE

Sex, Sex, Sex

THE JACK-IN-THE-PULPIT IS CONSIDERING a sex change. The violets have a secret. The dandelion is smug. The daffodils are obsessive. The orchid is *finally* satisfied, having produced over a million seeds. The bellflower is *not* satisfied and is slowly bending its stigma in order to reach its own pollen. The pansies wait expectantly, their vulviform faces lifted to the sky. The evening primrose is interested in one thing and one thing only.

A stroll through the garden is almost embarrassing.

ABOUT 80 PERCENT OF FLOWERS are hermaphrodites, both male and female. Pollination is the movement of pollen from an anther to a stigma. Fertilization occurs when the sperm from a pollen grain unites with an egg in the ovary.

Hermaphrodite flowers could easily pollinate and fertilize themselves. Most don't. Instead, they try to mix and match their pollen and eggs with the pollen and eggs from flowers of the same species: I'll take this. You take that. Here. Yes.

Sex, good sex, is all about cross-fertilization. Why?

Why have sex at all?

In terms of the individual and its offspring, asexual reproduction is so much easier. You don't have to think about males or male parts. You cut your investment in half. You don't have to use up all that energy and time. You just reproduce.

In a sexual population, an asexual mutant has many advantages and should quickly spread and take over. In an asexual population, a sexual mutant has many disadvantages and should quickly die out.

Scientists are still puzzled by the question: What good is sex?

They have some theories.

In a cell, when genes divide and replicate themselves, their occasional changes or mistakes are sometimes harmful and even lethal. But when an individual gets a set of genes from two different parents, that dangerous mutation can be neutralized. The normal form of the gene usually takes over, and the mutation is not expressed. In the offspring of asexual reproduction, the harmful genes tend to accumulate.

The recombination of genes from two different parents also allows for diversity among offspring. For natural selection to work, the genetic recombination has to create an immediate advantage for those individuals. In a variable world, their variability may mean that more of them survive.

Finally, there is the theory of the long run. Natural selection would not favor sex or cross-fertilization because these things are good for the species. Natural selection does not care about the future of the species. But sex and cross-fertilization *are* good for the species because they prevent the buildup of harmful mutations and because they produce a population that is diverse. When the climate gets colder, when pollinators disappear, when new diseases attack, the population may have individuals that can survive and reproduce. In the long run, species lucky enough to be sexual—species that, for complex reasons, resist asexuality—may simply be the ones that last.

THESE ARE ONLY THEORIES. But you're convinced. You decide to be sexual. And you decide to cross-fertilize.

First, you must avoid clogging up your stigma with your own pollen grains.

Some flowers, like the delphinium, separate their sexual parts in time. In a version of cross-dressing, they go through a male stage, when their anthers produce pollen. Then, in a matter of hours or days, they go through a female stage, when the stigma is ready to receive pollen. In the passionflower, the stigmas curve down at this point, bending back to fit between their own anthers, closer now to the colored mosaic of petals, closer now to the pollinating bee.

A few flowers reverse the process, stigma first, anthers second.

Flowers also separate their parts in space. In many flowers, the stigma rises well above the encircling stamens. An insect first plops on the stigma as a good place to land, deposits its pollen, and then goes exploring, rummaging around the petals, and collecting new pollen. In the rockrose, the anthers are sensitive to touch. Once a pollinator has visited the flower, the anthers splay down and away from the central stigma.

The position of these organs is never casual.

Some plants have two sexes, much like animals. The willow has a male form with flowers that only have stamens and a female form with flowers that only have stigmas. There is a Mr. and Mrs. Mistletoe, a Mr. and Mrs. Stinging Nettle, a Mr. and Mrs. Cottonwood, and a Mr. and Mrs. Holly. This is the most dramatic separation of parts.

Some plants have two sexes but on the same inflorescence. Other species mix up their inflorescences with hermaphrodite flowers, male flowers, and female flowers.

Plants juggle their sexual parts and move around their sexes as a way of avoiding self-pollination.

A few plants also have the ability to choose their sex. An individual bog myrtle will produce only female flowers one year and only male flowers the next. It's not indecision. The myrtle is responding to water or nutrients in the soil, to light, or to temperature. Commonly, female flowers require more resources and more time to produce fruit; in a difficult situation, a plant reasonably decides to be male.

A young jack-in-the-pulpit is often male in its first season. When it is bigger and stronger, when it has stored up a supply of starch, it will consider the more ambitious female lifestyle.

IN POLLINATION, a pollen grain lands on a sticky stigma. The grain absorbs moisture. The grain swells, cracks, and sprouts a pollen tube, which pierces the stigma and grows down the style. The tube contains two sperm cells that are delivered to the ovary.

In fertilization, a sperm cell fuses with an egg to become the embryo of the seed. The second sperm cell fuses with two other cells to become the endosperm, which feeds the embryo. This "double fertilization" means that the seed will have plenty of food and can mature quickly. Double fertilization has given flowers a tremendous advantage over nonflowering plants. It has given humanity an array of fruit and edible seeds: in short, agriculture.

A flower cannot always prevent self-pollination. Accidents happen. The wind blows the wrong way. A bee misbehaves. The deed is done.

At this point, some flowers can still prevent self-fertilization. In many grasses, when the stigma recognizes a pollen grain as too familiar, it blocks the growth of the tube. Evening primroses block the tube just below the stigma. Lilies and poppies cause the tube to burst further down the style. In the red skyrocket, the tube grows down the style, enters the ovule, and even fertilizes an egg—which is then aborted. Flowers like these are self-incompatible.

Some flowers are staunchly self-incompatible.

Others waffle.

In some species, pollen from another flower is simply given an advantage. Pollen tubes that result from cross-fertilization, for example, may grow more rapidly down the style. The race is not fixed, but the handicaps are heavy.

All self-incompatible systems leak to some degree. A number of flowers simply change their minds. Breeding with yourself is better than not breeding at all. At the last possible minute, unpollinated stigmas may bend down or around to contact their own anthers or to pick up pollen left on the style.

A few species have two different flowers, ones that cross-fertilize and ones that self-fertilize. In early spring, violets cover the woodland floor. Later, if a violet fails to get pollinated, the plant produces a second green bud that never opens and barely rises above the ground. Unseen, unnoticed, the bud fertilizes itself.

Most flowers self-fertilize as a backup plan. Then there are the habitual "selfers." These tend to be flowers that must bloom and die quickly in unpredictable or harsh environments. The flowers are often small with little color and scent. They may look juvenile or undeveloped.

Plants that habitually self-fertilize survive where other plants cannot. Since they reproduce quickly without pollinators, they often colonize new areas. Eventually they become genetically uniform, with fixed genes or traits. Populations of the same species that have followed different lines of self-fertilization are sometimes mistaken for different species. In the nineteenth century, one botanist saw two hundred species of grass in the forms of one tiny self-fertilizing plant.

Flowers can take self-fertilization a step further. In a dandelion, the ovaries set seed without the presence or benefit of male sperm. Those seeds are genetic copies of the mother only. Agamospermy (what botanist Peter Bernhardt calls "virgin birth") is found in many plant families. Oddly, some of these flowers still require pollination, which stimulates the ovary even though the pollen has no other role.

Dandelions do produce some pollen and seem to attract insects. Dandelions are not stupid. They also have a backup plan. About 1 percent of seeds in a dandelion head are from cross-pollination. The flower has not lost its ability to produce variable offspring in a changing world.

Backup, backup, backup.

Various plants reproduce vegetatively, sending out runners or roots that will grow into clones of the parent. The oldest known living plant is a clonal creosote bush in the

Mojave Desert. The bush started as a seed some twelve thousand years ago. As you might suspect, this ancient shrub has a backup plan. In the rainy season, it blossoms with small, yellow flowers.

A SINGLE FLOWER genus can show the range of sexual strategy. The large, showy lady's-smock *(Cardamine pratensis)* is cross-pollinated by many insects and is largely self-incompatible. The small bittercress *(C. amara)* is pollinated by flies and is easily capable of self-fertilization. The smaller hairy bittercress *(C. hirsuta)* is a habitual selfer.

Flowers are flexible. Flowers are determined.

The flower of a European orchid resembles the female of a certain bee. In parts of the Mediterranean, the flowers of a related species are grabbed by a lusty male bee and pollinated. But when that bee died out in western Europe, the orchid evolved into a habitual selfer. Now, a few days after the flower has opened, its pollinia (masses or sacs of pollen attached to a stem) lazily fall out of the anther, hang in front of the stigma, and wait for a breeze.

Give that orchid a pollinator and it would return to cross-breeding. If that pollinator looked like a helicopter instead of a bee, the flower would consider the situation.

We humans do as strange—or stranger—things for sex.

SIX

In the Heat of the Night

ONCE UPON A TIME IN LOS ANGELES, California, a man and a woman met in a public garden with a large display of flowering philodendrons. The glossy green *Philodendron selloum* is a well-known houseplant. But potted philodendrons rarely produce a flower. These outdoor philodendrons rose up in full bloom. The man and woman were struck by the same observation.

"It's very. . . . " the woman said.

"Yes," the man agreed.

The creamy white flower of *P. selloum* is a rod about nine to twelve inches long, an inch or so in diameter, shaped like a penis. The flower is really an inflorescence with hundreds of white florets, each about the size of an uncooked rice grain, growing on a common stalk, or spadix. The spadix has three kinds of tightly clustered flowers: fertile female flowers at the bottom, infertile male flowers in the middle, and fertile male flowers at the top. The spadix is cupped and enveloped by a longer leaflike bract, or spathe, green on the outside, with a light yellow interior.

The man and woman began a conversation that lasted the rest of their lives. It included a house and furniture and two children. One day the man died unexpectedly. For many years, the woman lived alone. When she was much older, she found herself walking through a neighborhood in Brazil, the native home of the philodendron.

Twilight darkened the air, which was scented with a faint, unfamiliar perfume. The temperature was 50 degrees Fahrenheit, and the woman had draped a light sweater about her shoulders. In front of another public garden, she stopped before a bed of flowering *P. selloum*. Their green spathes were loose, pulled away from the spadix. The woman bent down and touched a clublike rod.

It was hot!

She pulled back in surprise. She reached down again before sitting like a child, cross-legged on the sidewalk, in front of the flowers. The white spadix burned 115 degrees. The male florets generated heat. The heat vaporized a spicy, resinous smell.

Suddenly, on the sidewalk, the woman heard her husband whisper in her ear. She felt him touch her neck in the old way. Everything that had ever happened in her life was still happening now.

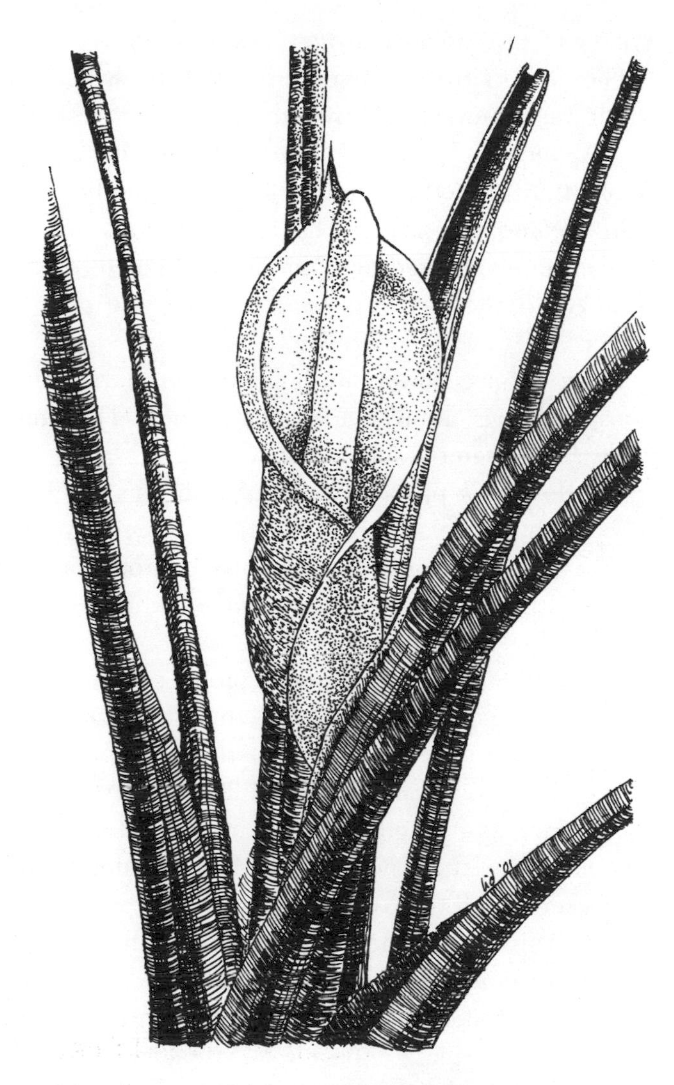

Philodendron selloum

WE HAVE LONG associated flowers with love.

The Greeks turned lovers into flowers. A beautiful youth was coveted by Zephyr, the god of the west wind, and Apollo, the sun god. Zephyr killed the boy, but Apollo transformed him into a hyacinth. Another young man became a narcissus. The anemone was once the hunter Adonis, adored by Aphrodite and killed by a wild boar. Aphrodite is the goddess of love; the rose, her flower.

Today we give flowers on holidays, birthdays, graduations, weddings, anniversaries, and funerals. The subtext is always the same: I love you.

Are flowers a physical form of love? It's a testable theory. Put aside your preconceptions. Imagine yourself fresh, new in the world. You are walking through a forest. You see a yellow columbine or a perfect white lily.

What do you feel?

In Brazil, *Philodendron selloum* produces flowers from the beginning of November to the middle of December, when the nights are chilly and one needs a light sweater. In the city of Botucatu, botanists have observed the inflorescence as it begins to heat up around dusk. The temperature of the spadix and the flower's odor peak between seven and ten o'clock at night.

At that time, too, dark beetles arrive, coming up from the soil or emerging from other philodendron flowers. The beetles follow the scent, zigzagging through the air until they come close enough for a visual cue. Then they fly directly into the spathe, crash, and fall to the bottom of the floral chamber, where the female flowers excrete a sticky, edible substance. In this safe, warm, dark place, the beetles crawl and feed and mate. As many as two hundred insects can fill a spathe, like ice cream in a cone.

After this period, the flower drops its temperature but still stays warmer than the cool night air. The female florets are becoming well pollinated by those insects which came here from other philodendrons. On the second evening, the fertile male florets release *their* pollen. The beetles swarm up the spadix, feed on pollen, are covered in pollen, and fly away to begin the cycle again.

Biologists get excited about the philodendron, which not only produces heat but also thermoregulates, raising and lowering its heat production in response to the outside air. In cold weather, the plant's thermostat is set at about 99 degrees Fahrenheit. The sterile male florets increase heat production if their temperature falls below that level and decrease production if the temperature rises. In hot weather, the florets stabilize closer to 115 degrees.

Warm-blooded animals thermoregulate by shivering and working their muscles. They increase their rates of breathing and blood circulation. This sends more oxygen and nutrients to the tissues, which need these things in order to generate more heat. Without heat, of course, we would die. Heat is what we carry in our blood. Heat is what we create. Heat is who we are.

Philodendrons also need oxygen and nutrients for heat production. But philodendrons cannot shiver. They get oxygen by diffusion, through pores on the florets of the spadix. They get nutrients from inside the sterile male florets, where fat droplets look surprisingly like "brown fat," a specialized heat-producing tissue found in mammals.

In an outside temperature of 50 degrees Fahrenheit, maintaining its temperature of 115 degrees, a *Philodendron selloum* creates as much thermal energy as a sleeping house cat. Roger Seymour, a zoologist who has done most of the recent research on philodendrons, likes to envision these flowers as "cats growing on stalks."

Roger became interested in the philodendron when he was handed a cutting at a dinner party. He was astonished to find it warm. He felt, perhaps, that he had been given some kind of enchanted animal, a creature turned into a plant, a cat under a spell.

You could say that Roger Seymour fell in love.

PHILODENDRONS BELONG TO the arum family, a large group that includes other heat-producing plants. In one species of lords-and-ladies (*Arum maculatum*), the spadix has four parts enclosed by the spathe: a cluster of female flowers, a bristly zone of sterile flowers, a cluster of male flowers, and a second zone of bristles at the top. The rest of the spadix, or appendix, protrudes from the spathe and generates heat and odor in the course of an afternoon. Some biologists refer to the appendix as an osmophore, or odor carrier.

The top bristles on the spadix act as a sieve, keeping out larger insects such as green bottle flies. Meanwhile, thousands of small midges are lured to the inflorescence, where they fall easily though the bristles to the bottom of the floral chamber. The chamber walls are coated in tiny droplets of oil. This slippery surface and the first zone of bristles keep the midges among the female flowers, where the insects feed on a sweet liquid.

Overnight, the male flowers open and rain down a golden shower. The midges, covered with the sticky fluid, are now also covered with pollen. The bristles wilt. The midges escape, only to be lured again by the odor of another flowering arum. Again they fall through the bristles. Again they pollinate the fertile females.

The voodoo lily is a tropical plant that can heat up to 27 degrees Fahrenheit above air temperature. For a few hours, on the first day of flowering, its warm appendix emits the smell of fresh feces, attracting flies and scavenger beetles. Later, inside the floral chamber, the base of the spadix heats up again for about twelve hours. The heat probably volatilizes the odor of starch-rich organs near the female florets. This sweet odor may trigger mating behavior among the insects, which stay inside the chamber until the male florets release their pollen.

Different lily arum species have different smells. One arum will remind you of apples, another of urine. One attracts carrion beetles and smells vile. Roger Seymour says it smells like a dead cat.

The flower of the skunk cabbage, also in the arum family, stays between 59 and 72 degrees Fahrenheit for two weeks in February and March. Botany books often show the plant surrounded by melting snow. It's a dramatic and puzzling picture since the cabbage's pollinators—beetles and flies—are not active this early in spring. The heat mechanism may be an evolutionary holdover, a habit passed down from skunk cabbage ancestors. Or, the mystery may simply require more botanists, sitting around in the snow, watching and waiting, in love with the skunk cabbage.

Thermoregulation in plants is not confined to a single family. It evolved separately in the sacred lotus, which can be 40 degrees Fahrenheit hotter than ambient air temperature. The Egyptians believed that the sacred lotus was the first living thing to appear on earth. When its petals unfurled, the Supreme God was revealed.

The inflorescences of some palms and the male cones of cycads (ferny plants that resemble palms) also produce a weak heat.

Roger Seymour, for one, is tremendously impressed. "The philodendron," he says, "produces more heat than a flying bird! It regulates its temperature with greater precision than some mammals!"

A sense of wonder is not only our starting point. It can also be our destination.

WHAT DOES A FLOWER have to do with love?

The ancient Greeks thought there was a connection. They made up stories which we read today.

The woman in Brazil, sitting on the sidewalk, thinks there is a connection. She counts herself lucky: her love made tangible, the flower burning, its desire hot.

SEVEN

Dirty Tricks

WHERE I LIVE IN southwestern New Mexico, yucca stalks of creamy white flowers bloom during the summer rainy season. These plants blaze up suddenly like candles in the desert, twelve feet high. Overnight, a scrubby field turns into a menorah.

Most yucca species are odorless, although many secrete a bit of nectar at the base of the ovary. This nectar may date from a time before the coevolution of yucca and yucca moth. For the yucca has only one pollinator and, on reaching adulthood, that ascetic insect does not eat or drink at all.

Breaking out of their cocoons, rising from the ground, yucca moths copulate in the white, waxy yucca flowers. The pattern varies among species, but typically a fertilized female moth climbs up a stamen in a flower and bends her head over the top of the anther. Uncoiling her tongue, she uses it to steady herself. With special mouthparts, she scrapes away the pollen and holds it fast in her forelegs, collecting pollen from as many as four stamens.

Now the yucca moth flies away to another flower in another yucca. There she pushes through the stamens and pierces the ovary, where she lays an egg. Then she moves up the tubular stigmas to pollinate the flower, pushing down pollen into the stigmatic duct. Often she lays another egg. After each one, she may move up again, rocking back and forth, pollinating the stigma, laying another egg, pollinating the stigma.

Yuccas are self-incompatible. By going to another plant to deposit her pollen, yucca moths cross-fertilize the yucca. By pushing pollen into the stigma, yucca moths are among the few insects that are active, rather than passive, pollinators.

In this way, they ensure food for their larvae. An unfertilized yucca flower soon falls from the plant. The ovules of a fertilized flower produce seeds. The yucca moth larvae hatch in the ovary and consume up to 15 percent of these seeds. Nourished, the larvae chew a hole through the fruit wall, drop to the soil, and stay cocooned until they emerge in a year or two or three. Enough seeds remain for the yucca to propagate.

All this sounds like an Aesop's fable in which the protagonists behave unusually well. When we think of flowers and pollinators, we often think of this kind of relationship. It is called mutualism: The butterfly feeds on the honeysuckle in exchange for transporting pollen.

Yucca Flower

Two different species have evolved to depend on each other.

Most mutualisms are generalized. Pollinators visit many plant species, and plants rely on many different pollinators. A one to one relationship, like the yucca and yucca moth, is less common.

Charles Darwin wrote, "Natural selection cannot possibly produce any modification in any one species exclusively for the good of another species; though through out nature one species incessantly takes advantage of, and profits by, the structures of another."

In the case of the yucca, natural selection has created an exceptional partnership.

It seems a model of cooperation.

It is almost a parable.

In their incessant way, other species quickly take advantage. Closely related to the yucca moth, the bogus yucca moth never transports pollen but also flies to a yucca flower, where it lays its eggs, which hatch into larvae that eat yucca seeds. This moth lays her eggs not just in new flowers but in already fertilized and developing ovaries. Not only does the moth fail to pollinate the plant, but fruits from these flowers will now have too many larvae, with too many seeds eaten.

Yuccas respond by shedding flowers (before they can seed and fruit) that are overinfested with insect eggs. Flowers insufficiently pollinated also tend to be aborted. In the end, the bogus yucca moth, which tries to cheat the system, won't reproduce as often as the moth that plays fair.

Botanists use the Spanish word *aprovechado,* "one who takes advantage," for an animal that benefits from a mutualism without providing any return benefit. All pollination systems are vulnerable to opportunistic *aprovechados*.

Moreover, pollinators themselves can turn into *aprovechados.* Instead of approaching a flower from the front, contacting the pollen-laden anthers, honeybees sometimes enter from behind and steal nectar by inserting their tongues between the sepals and petals. In the biologist's vernacular for crime, this nectar theft is known as "base working."

In flowers whose corolla has fused into a tube, thieving insects must forcibly bite through the tissue. Short-tongued bumblebees are notorious for using their mandibles to pierce the corollas of toadflax, daffodils, and columbines. Distinct from theft, nectar robbing is a more aggressive breaking and entering, in which the flower is damaged. Secondary thieves now enter the hole to steal nectar for themselves.

This is not a world of trust. Windows must be barred, doors locked. Flowers do what they can to protect themselves. Some have a leathery, hard-to-pierce calyx. Firm, overlapping leaves or bracts at the base also deter robbers. Dense inflorescences are another strategy.

Because the yucca and the yucca moth have evolved so tightly to fit each other's needs, they are particularly susceptible to *aprovechados*. They suffer, as well, from codependency. The reproductive success of one species depends on the reproductive success of the other. When farmers in the desert spray for agricultural pests, killing the yucca moths as well, the yuccas are left without a pollinator.

Statuesque, in full bloom, they light up the horizon. Serenely, they wait. No one comes.

This is botanical Shakespeare.

As well as using cross-fertilization, some yucca species propagate vegetatively, cloning themselves in a lonely fashion. It's a fall-back plan, a secret bank account, something your lover cannot possibly imagine, something you somehow forgot to tell him.

BECAUSE PLANTS DON'T MOVE, we think of them as nicer than animals. This is pure prejudice. As one researcher wrote, "cheating pollinators seem to be more rare than cheating plants."

Many flowers have the bad habit of exaggerating their virtues. They may have bushy hairs on their stamens or a bright yellow coloring so that the stamens look richer in pollen than they really are. Small anthers top large, attractive filaments that resemble anthers. In some plants, the sterile part of an anther puffs up into a nutritious-looking "dummy."

Flowers engage in what can only be called rough sex. In one orchid, a gentle touch on a tiny flower part causes the stem of the pollinium (a sac of pollen grains) to snap like a metal spring, catapulting the stem and its disk of sticky pollen onto the surprised bee. Sometimes the bee is knocked out of the flower. If a mischievous human tried this with a pencil point, the pollinium would sail for almost a yard. Other flowers similarly slam, hurl, or slap pollen onto an insect's body.

The ejection of pollen can be violent. Its placement can also be unfortunate. One orchid attaches its pollinium—the stem and disk—to a hawkmoth's eye. Although the pollen is transferred to another orchid, the relatively large stem remains glued in place. Proportionally, this would be like having a hockey stick hang from your eyeball. Some pollinators, such as birds and hawkmoths, can be seen with multiple pollinia studding their tongues. Darwin theorized that these animals, unable to feed, would soon die. First, though, they could pollinate a few more flowers.

Even the "nicest" flower can turn suddenly ruthless. In the common milkweed, pollen sticks so persistently to the visiting bumblebee that sometimes, as the bee tries to fly away, its entangled legs tear off.

A surprising number of flowers don't have anything to offer an insect but aggravation. As many as one-third of orchid species rely on deceit. Some specialize in pseudocopulation. Some look like a safe place to breed. Many smell like a food source, but instead of the promised reward, these flowers have invested in a dizzying array of chutes, passages, turnstiles, chambers, entrances, and exits.

Beloved by gardeners around the world, orchids have all the ambiance of a carnival fun house.

Drawn to one blossom by its pleasing stench, a fly settles on a tongue-like lip only to find itself flipped backward and downward, embraced by two springy "arms," and held fast. The scene starts to resemble a James Bond movie. The lip is hinged and balanced to respond to the fly's weight, and the two flower arms force the fly to struggle, removing any pollinium on its abdomen. Eventually, like James Bond, the fly escapes.

In the European lady's slipper, a fruity scent and a bright yellow color attract bees through an opening into the slipper or lip area. Large bees can usually exit, although a few insects get trapped and die. A small bee cannot fly out again and keeps sliding on the smooth, turned-in petals. After buzzing its wings and casting about, the bee finds a passage through the back of the flower, guided there by the light from translucent windows near the lip's base. As the insect squeezes past the stigma and stamen, any pollen it carries is left behind. New pollen is smeared onto its thorax.

The lady's slipper's plan is not foolproof. Some insects get away without having been properly accessorized. Also, experienced bees will avoid the flower. Prudently, the lady's slipper sends out rhizomes, underground stems that can root at a distance and produce a new, clonal plant.

In yet another orchid, one that rewards the euglossine bee with perfume, the big, drop-dead gorgeous flower hangs down and exhales an attractive scent. Part of its lip forms a bucket filled with liquid secreted by the flower. The base of the lip is slippery. The visiting bee loses its foothold and falls into the tiny swimming hole. Again, the exit is a hidden tunnel that passes the orchid's stigma and anthers. Here a bee might be caught for as long as thirty minutes, while pollinia are transferred to the base of its abdomen.

A few flowers don't bother to provide exits. The gnats that visit a certain arum lily in its female stage, when the stigma is receptive, will die in its lower chamber. If the gnats are carrying pollen from another arum lily, they will not die in vain. The flower has been fertilized. Gnats that visit the arum in its male stage, when pollen is being released, will find an exit and go free, nicely dusted.

The common jack-in-the-pulpit has male flowers and female flowers on two separate plants. Lured by the odor of fresh fungi, gnats fly to the flower and tumble into its chamber. Lucky gnats tumble into a male flower, which provides an escape. Unlucky gnats tumble into a female flower.

A particularly ugly scene takes place on a large, sweetly scented water lily. In the male stage, the lily offers an abundance of pollen-covered stamens to a host of hoverflies, bees, and beetles. For three or four days, each morning, the flower opens to offer a banquet Roman in scope and delight. Humans also take pleasure in the water lily. Many flowers are known for their beauty. But this flower seems to have reached a kind of Ur-beauty, an ascension to Buddha-hood.

In the female stage, the same lily opens and looks somewhat different. Now the stamens are without pollen and encircle a pool of liquid in the flower's center. At the bottom of the pool is the flower's flat, round stigma.

The background music changes.

We know what this means. We want to warn that little hoverfly: Don't land on that stamen!

Oblivious, the insect teeters on the now-smooth stamen surface. The hoverfly slips and tumbles into the pool.

Desperately the victim struggles, but the tall stamens offer no foothold. The liquid contains a wetting agent that clings and pulls at the lightest insect. The fly sinks beneath the liquid and drowns. The pollen on its body washes away, gently settling over the implacable stigma.

Sometimes, I am even touched by those carrion feeders trying to lay eggs in flowers that resemble rotting meat. They are clearly trying to do the right thing, breeding in a place that will provide food for their young.

Fungus gnats are similarly poignant. In flowers that mimic fungi, the parents leave satisfied, covered in pollen. The eggs hatch into larvae that will die of starvation. Some flowers try to hurry things along. In American wild ginger, the tissues of the flower are extremely poisonous.

PLANTS AND POLLINATORS are part of a mutualism more like an arms race than a marriage. In order to get more food or lay more eggs, the pollinator develops a new strategy. The plant counters. One side builds a missile. The other side builds an antimissile—and a bigger bomb.

The generation that wins threatens to destroy the system that supports them both. The water lily cannot afford to murder too many hoverflies, or the jack-in-the-pulpit too many gnats. The deceitful orchid must not be too successful, or its pollinators will starve to death. On their part, bumblebees and moths and hummingbirds had better not steal or cheat with such impunity that they fail to pollinate any flowers at all.

Missiles. Defenses. Antiaircraft. Images of battle come easily when you remember that many animals like to eat plants, not just sip nectar or collect pollen. Plants and insects have been at war forever, and pollination systems probably evolved out of that dynamic. Feeding beetles somehow shifted into pollinating beetles.

In the end, of course, when a flower cannot co-opt or avoid or out-strategize an enemy, it may have to destroy him.

The army worm feeds on daisies. The daisy's defense is to produce a chemical mildly toxic in the dark and highly toxic in ultraviolet light. As the worm eats the plant, it absorbs this chemical which eventually moves through its circulation system to the surface of its skin. On a nice spring day, the sun shines warmly. First the army worm glows florescent blue. Then it shrivels up and turns black.

A caterpillar known as the leaf roller has a counter-defense. It bends the petals of the daisy over itself and secures them with threads of silk. Safe in the dark, protected from sunlight, the leaf roller begins to eat.

Some plants go so far as to collude with their enemy's enemy. When spider mites start feeding on a lima bean plant, the plant emits a blend of volatiles. These chemicals attract another species of spider mite, a carnivorous one that eats the first population.

Alliances make for strange bedfellows.

Ants steal nectar from flowers. But most ants have a natural disinfectant that kills pollen sperm; typically, ants are not good pollinators. Plants sometimes respond by erecting barriers between the ground and the flower, sticky zones on the upper part of a stem or pools of dew that surround a stem and stop climbing insects like ants.

Plants also offer decoy nectaries, set away from the flower. In some flowering species, these supplies of nectar are exchanged for a guard of stinging, biting ants that protect the flower from other egg-laying insects or from corolla-piercing bumblebees. The flowers must also protect themselves chemically from their own guard ants, which are still thieves and cannot be allowed too close.

This is another parable of mutualism. Or another example of criminal racketeering.

We will find a story and a meaning in this because we are human and because story and meaning infuse our lives, just as scent infuses the life of the bumblebee.

IN THE EIGHTEENTH century, mutualism *was* a parable, of the perfect adaptations created by God. In a divinely balanced Nature, each creature had an unchanging role. All the parts of Nature worked together harmoniously, cooperatively, much as the parts of a human system should work together, each in his or her place, from peasant to king.

Science often reflects society, and we often want society to reflect what we know about the natural world. In the nineteenth century, the Industrial Revolution and new ideas of capitalism emphasized struggle in a healthy economy. Socialism and communism were counter-responses—a belief in mutual and shared power. Today we still seesaw between these two ideas, in politics and in botany.

Cooperation is nature's basic organizing principle.

Competition is nature's basic organizing principle.

In the environmental movement of the 1970s, biologists seemed to shift toward the former. They have shifted now toward the latter.

One of these biologists recently wrote:

> Plant and animal pollinators are mutualists, each benefiting from the other's presence. But the mutualism is neither symmetrical nor cooperative. Indeed, pollination derives evolutionarily from relationships that were fully antagonistic. The goals of plant and animal pollinators remain distinct—in most cases reproduction on one hand and food gathering on the other—and this leads to conflict of interest rather than cooperation.

Like sailors looking for land, scientists look for organizing principles. Physicists call it the grand unified theory. We are all looking for a grand unified theory. We all want the story that will organize the confusing parts of life.

The idea that function includes beauty astonishes me.

The idea that beauty includes violence knocks me for another loop.

EIGHT

Time

WE ARE DESPERATE TO understand time. How can the past be over?

Where is over?

We distrust time, the way it bends when you are eight years old, the way it slows when you are skiing down a hill. We know this is our fault, our oddball consciousness. Time is objective. We are not. The clock ticks.

Time waits for no one.

Then a physicist explains: Time is not separate from or independent of space. Moreover, space-time is curved or warped by the distribution of mass and energy. Time runs more slowly near a massive body like the earth. Put a very accurate clock at the bottom of a tower and another one at the top. The clock at the bottom will show a slightly different, earlier time than the clock at the top.

Take a pair of twins. Make one of them live in San Diego at sea level. Make the other live on a high mountain in Peru. The twin in Peru will age faster. Send one twin off in a spaceship traveling near the speed of light. Now things get complicated. When this twin returns, she will be younger than the one who stayed home.

Time can be influenced.

I AM PART OF A SUPPER CLUB that has lasted almost fifteen years, which seems a long time in this culture. Five couples meet every eight weeks. We each bring a dish from a selected cuisine, Chinese or Italian or Greek, and we all sit together for a few hours to eat enormous, delicious meals without our children, with a tablecloth, with wine. Perhaps we put a red rose in a vase.

The couples change. In fifteen years, some have dropped out of the group, some have moved away, some have divorced. Many of us see each other only at this dinner party. Among the things we value is continuum.

One night, before dessert, a couple received a phone call and announced suddenly that they had to go. Their cereus cactus was in bloom. It did not seem, at the time, a very good reason to leave a friendly group of people who had arranged with some difficulty this special occasion. We hid our annoyance, which we also filed, for future reference.

The cereus cactus is a slender, grayish plant often hidden under shrubby trees like mesquite or creosote. The stem can be as small as half an inch in diameter and grow as long as six feet. It is a spiny, unprepossessing twig. Botanists delight in scorning this cactus as an ugly duckling, with "the cuddle appeal of a dead stick."

Then, suppressing a smile, the botanists turn their backs. They turn around again. Giddily, gaily, they gesture with their hands. The cereus cactus has blossomed into a swan!

Its beauty, always, is a surprise. Its beauty, always, is a fairy tale.

The large, white flowers of the cereus cactus open at night into a silky, many-petaled star, with a musky, sweet smell. The star fills the palm of your hand, where it seems to glow in the darkness. On first seeing a patch of these flowers, one man believed that someone had turned on a dozen flashlights and left them in the desert, scattered under bushes, wasting batteries. One woman saw ghosts.

The Spanish name for the cereus cactus, *la reina de la noche,* means "the queen of the night." Each flower blooms once, for a matter of hours.

I have never seen a cereus cactus bloom.

My friend, the woman who left the dinner party, says it reminds her of time-lapse photography, the way she could see the flower unfold, the way it moved like royalty down a long carpet, gracious, inexorable. Her husband, whom I questioned later at another party, used the word "amazing" three times in a sentence. Once I accosted her teenage son on the streets of Silver City. He also nodded and said, "Yeah."

Cereus Cactus

My friend is a historian, director of the Silver City Museum. She tells me that in the 1870s, in this small mining town, people used to have cereus cactus parties. When their houseplant began to bloom, they would send out the news. Other people rushed over. Refreshments were served. Sometimes the event appeared in the local paper: "An Impromptu Soiree Enjoyed by All."

My friend laments that the annual flowering of the cereus cactus no longer means much to her husband and son. "Come see!" she exclaims. But they have already seen. They do not need to see again. "Neat," her son will offer as a gift. She has begun calling up friends instead.

Call me, I say.

The flower of the cereus cactus is ephemeral. Most flowers are ephemeral. Their lives are brief.

More than a few, like the queen of the night, live for only a day or an evening. In those hours, the silky blossoms of the cereus cactus must do everything they can to attract pollinators, night-flying insects like the white-lined sphinx moth, which come to drink the flower's nectar.

La reina de la noche cannot self-fertilize. Nor is she gregarious. As few as five to ten cacti may grow in an acre of hot, dry, inhospitable desert, mutely surviving the sun, the wind, the cows.

The queen compensates by being intense. Her fragrance is strong. Her beauty is legendary.

She has only this night.

Well, yes and no. The cereus cactus may have only a

few flowers, each flower lasting only one night, with the entire season lasting as little as four nights, depending on moisture. But the cactus itself can live to be seventy-five years old. The long, spiny twig will keep blooming, summer after summer, waiting for the right white-lined sphinx moth.

The queen in the fairy tale will sleep for years, while kingdoms fall and the prince hacks his way through a briar-filled woods.

IT MAKES YOU THINK about our allotment of time.

A clonal creosote can live to be twelve thousand years old. A redwood tree saw the Spanish missionaries. A human, a parrot, and a cereus cactus reach the Biblical age of three score and ten. A black bear may forage for thirty years. A dog may be your companion for fifteen. Mice rarely see their second birthday. Many insects don't last a month. No doubt, there are good reasons for each span of life, for the old age of a tortoise or the brief flare of a moth.

From a plant's point of view, flowers are brief because flowers are so much work. Beauty is expensive to maintain, all that scent and color, all that waving in the wind. The materials of reproduction are also fragile. They must be constantly protected.

In some climates, flowers have to think about weather. Winter is near. The rains are coming. It's getting hot. I'm getting dry. I'm losing all my petals. I'm freezing. I'm blowing away.

Fertilization may be the sine qua non, but better sooner than later.

Flowers that last a short time are often on a plant that blooms for much longer. In the morning glory, each individual flower opens and dies and is replaced overnight by another flower. The giant water lily blossoms until its pool is dry, but each extravagant lily lives for only two days.

Some flowering plants themselves have a very short time to live. Many annual wildflowers must sprout, grow, flower, set seed, and die in a matter of weeks. They often self-fertilize.

A few flowers manage to surprise us. Magnolia flowers live as many as twelve days. Orchids are among the longest-living flowers. Sheltered in a greenhouse, one orchid from Asia can stay fresh for nine months.

Flowers that live a long time tend to look sturdy. Their petals feel thick and waxy with a coating that retains moisture. Their veins may be fibrous, forming an internal skeleton that helps support the flower's shape.

Flowers pollinated by birds with sharp beaks or by beetles with chewing mouthparts need to be especially resilient, and these reinforcements add to the flower's life span. Flowers that have only a few pollinators may need to stay open long enough to get someone's attention. Flowers that practice deceit need enough time to attract one naive customer. Flowers that are firmly self-incompatible need additional time to cross-fertilize.

THE CENTURY PLANT, OR AGAVE, grows in the same desert as the cereus cactus. Depending on the species, the century plant does not flower for its first five years or ten years or fifty years. Then, when you have completely lost interest, the spine-tipped leaf rosette sends out a flower stalk like a huge asparagus. This stalk may grow as much as a foot a day to as high as thirty feet. Branches extend horizontally from the top. Buds swell. Masses of yellow, tubular flowers open at night, positioned high like the lights on a baseball field, smelling strongly of musk and rotting meat. The century plant has produced a flare gun for migrating bats, hummingbirds, and other pollinators that descend in a rush, feed, and leave.

Meanwhile, the rosette withers and dies. All its stores of food and water have gone into this growth of stalk and flower. The century plant does not live to be a century. It dies when it blooms, and it blooms once, staking everything on one shake of the dice.

Other plants also begin to die as soon as they flower. Commercial bedding annuals like marigolds and zinnias eventually convert all their leafy stems into flowering stems. These annuals hold nothing back until every growing stem is a flower, unable to make a leaf or keep the plant alive. (Many wildflowers keep some stems underground in the form of bulbs, tubers, or rhizomes. Unlike the zinnia, the same plant will be there next spring.)

Some plants do not die but do bloom only briefly, in a fireworks display. They are also high rollers and flashy dressers, ready to dance until they drop.

A plant like the morning glory is more circumspect. It has a long flowering season that is carefully measured out: here, here, here, that's enough, stop, come back tomorrow. This works for pollinators that live in the area, are long-lived, and have good memories.

In the U.S. Southwest, eight thousand years ago and for thousands of years after that, people tended fields of century plants, which they harvested for the inner heart and young flower stalk. The leaves provided fiber. The roots made soap. Many of these fields, bordered in stone, are still being found by archaeologists. The century plant is one of our first agricultural crops.

It is a crop still. Just before blooming—as the plant gathers sugar and nutrients, as it prepares for its first, final, and fatal puberty—the heart of the agave can be cut out, roasted, and beaten into a pulp that ferments into liquor. Our tequila comes from commercial agave farms. Each year, over a million wild agave plants in Mexico and the United States are also cut open for bootleg mescal.

We lift our glasses to the high roller.

PHYSICISTS AGREE THAT time changes when you are a little tipsy. When you have had some wine, when your friends are nearby, you look at a flower, a wilting rose, and see how fast it is moving, close to the speed of light. You see how big it is, so near its death. You see how it curves the darkness of space, how it warps the flow of time.

Time can be influenced. Time slows down when we look at a flower. Perhaps we age more slowly, too.

It's worth a party. Send out the news. An impromptu soiree!

NINE

Travelin' Man

Pollen has itchy feet. Pollen has a job to do, going down that long, lonesome highway, bound to leave, bound for glory. You can't hold him back. Hit the road, Jack. Pollen is a travelin' man.

The hazel tree sighs, a puff of pollen. In the first mild weather of spring, the male catkins of the hazel tree hang down like lambs' tails, their tiny flowers ribboned, two by two. When the wind blows, the catkins bobble. A cloud of yellow colors the air.

A fine, masculine dust.

On separate catkins, the female flowers of the hazel tree barely reveal their crimson tips. Well-bred, aloof, they will reject any pollen from the parent tree.

The yellow cloud drifts, casting its pollen shadow.

In an otherwise prosaic research paper, I find this sentence, which I am tempted to write as a poem, with the appropriate line breaks:

A paternal plant
spreads a thin
and particulate
sheet of itself
over the habitat.

A single catkin on a hazel tree can contain four million grains of pollen. The tree may produce several thousand catkins. The light, dry grains are designed for flight, to be cast into the breeze and lifted as high as nineteen thousand feet or borne as far as three thousand miles. It's a wild ride and often pointless. Most of the pollen will fall back to the earth to dry out in the sun or to drown in a pond or to dust the wrong plant.

A few will swell and rupture in the mucous membrane of a human nose, triggering the defense of our immune system: Help, help! A foreign body, more fluid! More fluid over here!

The result is an eruption of tears and sneezing.

A pollen shadow is cast over the land. One by one, each grain is accounted for.

This is where the hazel tree uses math. Four million grains of pollen multiplied by several thousand catkins increases the chances that some of these will land on a compatible stigma.

The majority of plant species in the world depend on the discriminating animal pollinator. But in terms of biomass—the plants that cover the earth's land—most plants cast their sperm into the air. For the dominant trees of a forest canopy, for conifers and pines, for grasses and sedges and rushes, wind pollination is the most effective choice. Even swarms of insects could not handle this job. In areas where insects and birds are scarce, such as salt marshes and some deserts, flowers also rely on the wind.

Flowers judge, as best they can, how and when to release their pollen. To avoid storms, wind-pollinated plants tend to bloom in the milder days of early spring and autumn. Similarly, grass flowers may open early in the morning or late in the afternoon, when the turbulence caused by heat won't take their pollen into the next state. In a dead calm, grass flowers prevent their pollen's release by holding their grains in the spoon-shaped lower end of the anthers.

On the very nicest of days, when the air is slightly fresh, when the sun is pleasantly warm, when troublesome insects have not yet appeared or are already gone, we live and breathe in an effluvia of male sex cells.

POLLEN GRAINS VARY in their tininess. The forget-me-not's is three microns (three thousandths of a millimeter). Pollen from a pumpkin can be eighty times larger than that, visible to the human eye. Most species of plants have pollen grains about thirty microns long.

Each pollen grain is enclosed by a sturdy outer shell that may be sculpted variously with spines, warts, or rounded or angular ridges. These patterns are unique to groups of plants and sometimes to a single species. Wind-pollinated grains tend to be relatively smooth and aerodynamically efficient. The most elaborate surfaces, with scary protrusions and medieval spikes, are found in flowers pollinated by insects, the better to catch and latch onto a thorax.

In animal-pollinated flowers, the pollen we see is usually a clump or sac of smaller grains held together by a gluey cement that adheres easily to the beak of a bird or the shell of a beetle. The adhesive oils produced by the grain contain pigments that cause pollen to appear yellow, orange, green, blue, black, or brown. The colors may attract pollinators. The oils may give off scent, serve as a water-repellent, or protect the grain from ultraviolet light.

Pollen leaves an anther in various ways. Often the anthers naturally dry and split along preformed seams. The drying process can be a quiet one or so melodramatic that the stamen jerks and curls. At the slightest touch, the anthers in some orchids machine-gun-spray their pollen: bang, bang, bang.

Flowers often protect their pollen. Some anthers have the ability to close again if the environment becomes too wet or cold. The anther cones of certain flowers contain pollen grains that can only be released through pores at

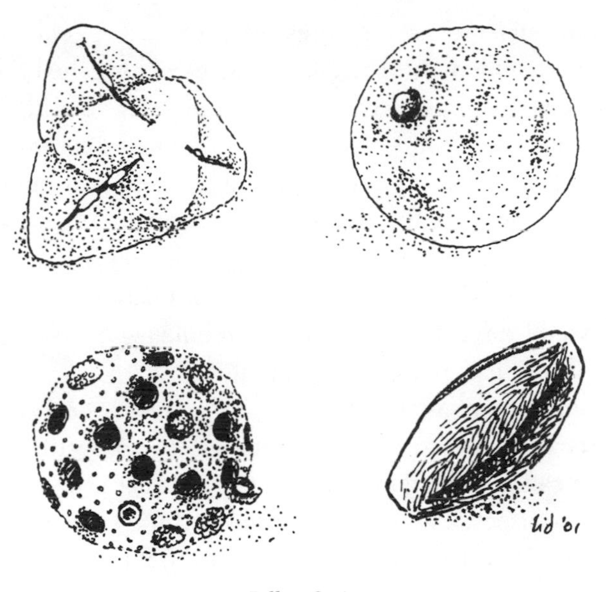

Pollen Grains

the anther's tip. This also keeps the pollen safe and dry, until the right suitor comes along. A bee lands on the anther and vibrates its thoracic muscles in the frequency needed to free the pollen. The wrong vibration fails to produce pollen or produces fairly little. Honeybees do not seem to vibrate very well. Instead, they do strange and useless things, like trying to stick their tongues into the anther pore. Bumblebees have better buzz. About 8 percent of the world's flowers, including tomatoes, potatoes, blueberries, and cranberries, require a bumblebee at the window, serenading the pollen out.

Many insects have bodies adapted to get, eat, and transport pollen. More than most, bees have reshaped themselves into pollen porters. The hind leg of a honeybee worker is a kind of Swiss Army knife. It includes a pollen basket, or concave region with hairs that anchor pollen; a pollen rake, or row of stiff bristles; a pollen press, or flattened area; and more pollen combs, or rows of stiff hairs. These parts work in coordination as pollen is passed from the forelegs to the middle legs to the hind legs, where it is packed into a pellet.

Pollinators claw, scratch, pry, grab, press, and pack. On its part, pollen is hardly passive. In some flowers, pollen can actually jump across the gap that separates it from an insect. The force of desire is static electricity.

Plants have their own electrostatic fields, which are strongest on clear, warm, sunlit days, with the "terminal" ends of the plant, the flowers, having the greatest charge. Specifically, the dry parts of a flower usually have a negative charge.

Bees just leaving a hive also tend to have a slight negative charge. But as they fly through the air, and as friction strips away electrons, the insects become positively charged. When the foraging bee nears a negatively charged dry anther, pollen grains can leap out and attach themselves to the insect's body.

Later, as a passenger on the bee, the pollen grain also becomes positively charged. It can now jump again, this time to a negatively charged stigma.

A sailor reaching land.

A pilot touching ground.

A traveler wearying of traveling.

Cold, lost, the wayfarer knocks at the cottage door. There is a light in the window. There is the smell of home.

If the pollen grain is lucky, this will be a compatible flower of the same species, not on the same inflorescence, and not too closely related, not a daughter or granddaughter of the pollen's parent plant.

The surface of every pollen shell, whether smooth or sculpted, has openings that allow the grain to release moisture and lose weight when it leaves the anther. These same holes can now absorb moisture, which will rehydrate the grain when it lands on a stigma.

Some stigmas, as in certain daisies, are dry. The outer cells of these stigmas first "read" and approve the identity and source of the pollen and then secrete the necessary liquids.

In many flowers, the receptive stigma is already wet. Here the grain sticks easily and absorbs sugar water from the stigma's surface. Soon the grain cracks, swells, and sprouts a tube. In some flowers, the pollen tube grows by "drilling" its way through the style's tissue. In other cases, the style already has a hollow channel or jelly-like areas easily penetrated.

Pollen from compatible flowers is welcome. Pollen from incompatible flowers usually goes off in the wrong direction, explodes, or stops growing.

In the meadow saffron, a tube will reach the ovule about twelve hours after pollination. In some flowers, the tube whistles down in six hours. At the ovule is an entrance pore into which the tube delivers two sperm cells, one for the endosperm, one for the egg cell.

This is the good news. In flowers pollinated by animals, this is what happens to "good" pollen grains.

THE BAD ONES, of course, get eaten.

Most pollen is sacrificial.

For many insects, pollen is breakfast, lunch, dinner, and snacks. To a bee going shopping, the nutrition label is promising: 16–30 percent protein, 1–10 percent fat, 1–7 percent starch, no sugar, various vitamins, various minerals.

Flowers can be extraordinarily generous. The corn poppy produces over 2½ million pollen grains per flower, overloading its visitors in the expectation that just a few of these grains will not get consumed, that a few, out of millions, will get taken to another corn poppy.

To entice customers, other flowers produce fake, sterile pollen, less costly to make, but almost as nutritious. Smaller hidden stamens, with fertile pollen, are positioned so that the insect can contact them easily.

For the honeybee, pollen is usually plentiful, easy to find, and labor-intensive. This is not convenience food. The bee shakes, rakes, presses, packs, and flies back to the bee colony. Here the pollen must be treated chemically so that it won't germinate, processed for storage, and made into "bee bread" to be eaten by larvae and adults. (Nectar, which becomes honey, is stored differently.) The hard, spiky walls of the grain are a problem. It can take three hours for a honeybee to digest a pollen mass.

ONCE ON A STIGMA, pollen promptly responds to moisture and other chemical signals. Without those signals, the outer rind of a pollen grain is so stable that it has been called the most resistant organic material known, a natural polymer as tough as industrial plastic. Although the inner potency of pollen is short-lived, the surface fiercely resists rot, pressure, and temperature extremes. Scientists have found pollen grains, eaten but not digested, whose walls have survived for thirty thousand years in the stomach of a frozen mammoth. Fossilized pollen can be much older.

Archaeologists, naturally, love pollen. So do paleontologists, climatologists, geologists, and forensic scientists.

The Neanderthals, for example, buried their dead with whole flowers some fifty thousand years ago. We know this because of the pollen grains left behind: ancient versions of blue hyacinth, yellow groundsel, knapweed, and yarrow. The idea that these people's love of flowers somehow makes them more real to us. Suddenly, we see them weeping. We see a belief in the afterlife. We see culture.

In 1994, a mass grave with the skeletons of thirty-two young men was uncovered in Magdeburg, Germany. The victims were either Germans killed by the Gestapo in the early spring of 1945, or they were Soviet soldiers killed by Soviet secret police for refusing to break up a German revolt in June 1953. The nasal cavities of seven skulls produced plantain, lime tree, and rye pollen, all heavily emitted in June. The murderers were Russian.

Traces of pollen also have been found on the Shroud of Turin, a linen cloth that bears the image of a wounded man, which some believe to be the burial shroud of Jesus Christ. Since 1538, the cloth has been enshrined in a cathedral in Italy. Pollen from species of bean caper and tumbleweed confirm that the material originally came from Israel.

Pollen keeps traveling.

Traditional Navajos believe that the "pollen path" is the way between the gods and humanity. It is the harmony that should exist between us.

> In the house of life I wander
> On the pollen path
> With a god of cloud I wander
> To a holy place
> With a god ahead I wander
> And a god behind
> In the house of life I wander
> On the pollen path

We are all walking the pollen path. We all breathe in that fine, masculine dust (and suffer, some of us, from our body's defensive, surprised reaction).

A paternal plant spreads a thin and particulate sheet of itself over the habitat.

Flowers everywhere catch the falling clouds. The crimson tips of the hazel tree hide, wait, and receive. The elaborate stigmas of grass comb the air. The corn poppy prepares a feast. The pollen grain leaps ecstatically from the bee.

One-half of something meets the other half.

Bound to leave, bound for glory.

TEN

Living Together

THE YOUNG PLANT IS paying close attention.

Photosensitive cells in its leaves and stems "see" across the spectrum of visible light, from far-red to ultraviolet. The plant knows it is day and not night. The plant knows that the days are getting longer. The plant is detecting a very hot day, with a lot of short, damaging ultraviolet wavelengths. In the plant, two genes are activated. They produce a colorless pigment, a sunscreen to filter out the harmful light.

The young plant is busy sending down roots that taste and test and forage for nutrients. When they reach an area rich in minerals or salts, these roots quickly, greedily, grow lateral rootlets to collect the food. Above ground, the plant also tastes and tests, absorbing chemicals from the air or from the bite of an insect. The plant feels the wind buffet its stalk. It responds with a growth of cells that strengthen those tissues. The plant tingles, sensitive to small electric currents. A thunderstorm is coming. Rain means more growth. There are preparations to make.

The big question remains: When to reproduce? At the right hormonal signal, the tip of a leafy shoot will stop adding leaves and start making a flower. Often these hormones are triggered not by light, but by the length of uninterrupted darkness, something being carefully measured by the plant's green leaves. Plants that bloom in the early spring or fall require a cycle of long nights and short days. Plants that bloom in the summer require a cycle of short nights and long days. Some plants need a second cue of temperature. They bud in the fall and bloom in the spring after months of cold weather. Some flowering plants, like tulips and hyacinths, are only cued by temperature. Some plants use rain as their stimulus to flower. Others wait for the dry season.

This young plant is waiting for its own burst of hormones, which it will produce when the cycle of darkness is absolutely perfect.

In all of this, in all of its short, sweet, individual life, the young plant is influenced by other plants. Most of them are competition. They eat the young plant's food, use its water, absorb its sunlight. The young plant has to respond quickly. If it were to sense now, for example, that it was not getting enough good light, it would begin to grow faster, taller, higher, struggling out of the shade of those nearby growing plants. If it were a sunflower, it would send out a toxic compound to inhibit the growth of that nearby dratted evening primrose.

This young plant is also a competitor.

In the end, surprisingly, some of these nearby plants may prove beneficial. Even now, the young plant has a relationship with a fungus that helps its roots absorb nutrients. (Fungi are not plants. But they are often closely associated with plants, as friends or enemies.) When this young plant has flowered, other flowers in the area may help attract pollinators or repel pests. Their flowers might be good models to mimic. They might have something to borrow or steal.

The young plant lives in a community of plants. We humans have a great attachment to the word *community*, which we have invested with a sense of nostalgia. We need community. We used to have more community. We grieve for community.

We have forgotten, perhaps, that community is not only supportive. Community also stones the adulterer and shames the unconventional. For the individual, community is a mixed blessing, both good and bad.

Community is your neighbor. What's he doing now?

FLOWERS CAN be good neighbors.

Red skyrockets next to blue delphiniums bloom sequentially, first one species, then another. This extends the period during which a pollinator can search for food. Some insects need that time in order to reach sexual maturity, reproduce, and start the next generation. Working together, overlapping, different flowers support the pollinators they need for their own reproduction.

Flowers that open at different hours of the day also allow pollinators to work all day long. Flowers with different rewards help sustain those pollinators: The bee collecting pollen for the hive might also need a drink of nectar, just to keep going.

Flowers communicate with their pollinators through scent or odor molecules released into the air. Plants also talk to nonpollinators, other insects. Often enough, they ask for help.

Parasitic wasps sting caterpillars into which the wasps lay eggs. As the wasp larvae grow, they feed on and kill their hosts. Naturally, caterpillars hide from wasps. In a large field of leafy plants, how do wasps find their prey?

When plants "taste" certain secretions from a caterpillar, they send out compounds into the air. Wasps recognize and follow these compounds. Quick, the plant says, I've got him. He doesn't suspect a thing. He's right here, under this leaf. He's *eating* this leaf. Better hurry.

In the 1980s, researchers used willows and maples to show that pest damage in one tree could lead to a greater resistance toward pests in neighboring trees. The damaged plants might have been sending out a chemical warning, allowing nearby plants to start their own defense. The "talking trees" studies were met with criticism and ridicule. Today, these scientists are being vindicated. In new and better-controlled experiments, scientists have shown clearly that plants attacked by

Delphinium

pests do send out wound signals. Nearby plants, not yet damaged, are then better able to repel these pests or to attract their predators. Perhaps the airborne chemicals are simply being absorbed and used by the undamaged plant. More likely, these chemicals activate genes and begin the plant's own response.

Most of this research has been done on crops. Normally, we wouldn't cheer on a wasp laying an egg in a paralyzed caterpillar—except when that caterpillar is infesting our corn. We are relieved to know that tomato plants can defend themselves aggressively. We are interested in cabbage and cabbage aphids, in lima beans and spider mites, in sugar beets and armyworm larvae.

As we learn more about plants, it is not hard to suspect similar defenses in a meadow of wildflowers: deep blue monkshood, purple-blue delphinium, sky blue flax, yellow columbine, golden sunflower, pale lousewort, elephant's head, shooting star, red skyrocket, Indian paintbrush. Perhaps these plants, too, are talking to each other.

A MEADOW, OF COURSE, is not really a suburban neighborhood. It's more like a shopping mall. Everyone is here to do business. Reasonably, in this mall, there can only be so many shoe stores, so many restaurants, so many boutiques. When two flower species are too similar, when they compete too directly, one of them, reasonably, should go out of business.

That doesn't seem to happen. More often, flowers find a way to adapt. Some shift to self-fertilization. Some start offering a different reward, perhaps an entirely new product, like resin or oil or perfume. Some change their flowering time. The red skyrocket grows near a certain penstemon. In different parts of the United States, either species might flower earlier or later, depending on what and how well the other one is doing.

Between specific plants, however, and between species, competition can be intense.

In the Southwest, the creosote bush and a small shrub called burro-weed share the desert's resources. The plants have become territorial, spaced well apart. A burro-weed root will stop growing if it enters into the root zone of a creosote bush. The creosote is emitting a growth inhibitor. Even the root of another creosote bush will be stopped, barred by the same chemicals in the soil.

Burro-weeds are less effective against invading creosote bushes. But when the root from one burro-weed touches the root of another, there is also a decline in growth. When roots from the same plant meet, nothing happens. The plant recognizes both a self and a nonself.

Allelopathy is a plant's bad magic, the release of substances that harm nearby plants. As early as 1 A.D., the Greek scientist Pliny saw that not much grew under the black walnut, its shade being "heavy" and "poisonous." Weeds like lamb's-quarter, thistle, nutgrass, and chickweed probably do not just compete for resources; they also affect the healthy growth of nearby vegetation. Many mustards and sunflowers are likely allelopathic, as are goldenrods and asters. In nature, pure stands of one kind of tree or grass might well suggest an enforced segregation: No trespassing. Get out. This means you.

Some plants actively prey on each other. The seeds of witchweed only germinate in the presence of cereals like sorghum, maize, and barley, as well as crops of tobacco and cowpeas. As these plants start their growth, the weed also grows rapidly underground, reaching with witchy fingers toward its victim and developing a specialized rootlike organ. This organ allows the parasite to suck out nutrients and water from the roots of the host plant. Eventually the witchweed appears above ground and produces a pretty, red flower.

By then, the farmer has probably lost her field of sorghum. In parts of Africa and Asia, witchweed can affect as much as 40 percent of arable land. These are places where a failed crop is a family disaster, where children die from hunger, where witchweed is as deadly as war.

Other parasitic plants, like cancer-root or mistletoe, attack hardwood trees. The sandalwood tree gets its food from nearby grasses. Indian pipe has a ghostly, white, tubular stem with a single white blossom. The plant lacks chlorophyll and feeds off the fungi that have become part of a nearby tree's root system.

In a meadow of wildflowers, among the delphinium and columbine, the strangest parasitism may be practiced by another kind of fungus, the rust fungus *Puccinia,* which infects mustard plants and reprograms them to grow in unnatural ways. An infected mustard plant looks very different from an uninfected one. The alien may have twice as many leaves and leaf rosettes and be half as tall. Elongated stems are crowned not by mustard flowers but by clusters of bright yellow petal-like leaves that exude a sticky, sugary substance. Bees, butterflies, and flies visit this pseudoflower. In the same way that they spread pollen, they spread the reproductive cells, or spores, of the fungus, which now need to find and combine with other fungal spores.

Importantly, the odor of the pseudoflower is different from the odor of the host plant's flower and vegetation. The scent is also different from concurrently blooming flowers such as buttercups and phlox. The pseudoflower produces a unique smell, which may encourage constancy, helping pollinators carry the fungal reproductive cells to another fungal fake, not to another flower.

From a distance, a botanist might mistake this counterfeit flower for a real one. Up close, the rust fungus *Puccinia* can fool you and me.

IN FLORAL COMPETITION, in that struggle to win, a different approach resembles aikido: Use your opponent's strength.

In Batesian mimicry, a flower tries to look like another species. The orchid resembles a lily that provides nectar.

The orchid does not provide nectar. This kind of deceit must be restrained. Too many mimics defeat themselves; their pollinators wise up and switch to another food source or don't wise up and die from lack of food. Batesian mimics depend on the abundance of their model. The more successful the model, the more successful the mimic.

Most Batesian mimics pretend to be a flower that offers a reward like nectar or pollen. Some flowers push the envelope. One orchid, bobbing in the breeze, tries to imitate the general movement of an insect so that a certain territorial bee will swoop in and buffet/pollinate the flower in an effort to drive away the offending insect/orchid. The bees do not always cooperate. The orchid also self-fertilizes.

Batesian mimicry is most often cited in animals, where the point is not to attract pollinators but avoid predators. The harmless kingsnake looks like the poisonous coral snake. An ugly, nontoxic caterpillar looks like an ugly, toxic caterpillar. In both instances, the resemblance to something else benefits the mimic and no one else.

Another kind of mimicry, called Mullerian, is rather different. Here, the resemblance is mutually beneficial.

A number of plant families include species with inflorescences of small, white florets. These umbellifers all have similar shapes and are visited by a variety of insects. They may be showing Mullerian mimicry, much like the convergence of so many yellow-centered white "daisies" or yellow-headed dandelions, hawkweeds, and their relatives. Sharing the same advertisements, borrowing from each other, favors the group as a whole by attracting more pollinators.

In the U.S. West, as many as nine species from seven different families have flowers that are red and tubular and bloom at the same time. Migratory hummingbirds like red, tubular flowers. Each flower species places its pollen on different parts of the bird's body, which allows each flower to find and fertilize a similar flower.

Eight of these species are combining their resources of nectar to support a larger number of customers. The ninth species practices Batesian mimicry. It is red and tubular and nectar-less.

At some point, reading botany, the nonbotanist raises her head to ask, "Who was Bates anyway? Who was Muller?"

Flashback to the Amazon River, 1848. Henry Walter Bates is traveling tippily in a canoe down a tributary, marveling at the monkeys, fish, and butterflies. Bates is twenty-three years old; his companion, Alfred Russel Wallace, twenty-five. They are English boys trying to make a living as naturalists and collectors.

Bates explored the Amazon basin for another ten years. In the end, his collection of insects included eight thousand new species.

One day, watching a swarm of South American butterflies, Bates recognized two different species, with the second species closely resembling the first. The first species is distasteful to predators. The second species tastes fine but uses its coloring to deceive predators. On his return to England, Bates read a paper about these butterflies before the Linnaean Society, the premier scientific group of his age.

Some years later, the Brazilian zoologist Fritz Muller described another kind of mimicry. Two bad-tasting species might begin to look like each other in order to maximize their protection from predators. Viceroy butterflies were once thought to be Batesian mimics of monarch butterflies. Actually, to birds, both insects taste bad. The butterflies have combined their strengths. The predator learns, doubly quick, to spit them out.

Bates's companion, Alfred Russel Wallace, also left the Amazon basin and continued collecting in Malaysia. What Wallace saw on those islands was similar to what Charles Darwin had seen in the Galapagos Islands, twenty years earlier. Excited, Wallace sent Darwin a letter. Darwin wrote back.

Finally, concerned that Wallace might publish his theories first, Darwin finished his long-overdue work on natural selection. In 1858, the two papers, written separately by the two men, were read before the Linnaean Society.

Darwin's *Origin of Species* was published within a year.

Soon after, Bates presented his work on butterflies. It seemed a perfect example of how natural selection worked. Individuals that resembled the toxic models were favored, not eaten by predators. More of them passed on their genes. More individuals looked more like the model. The species, as a whole, became a mimic. Darwin sent Bates an enthusiastic letter.

They began talking to each other.

SCIENCE AND FLOWERS have a few things in common. It's all about community: cooperating, competing, stealing, borrowing, exploiting, combining.

A field of wildflowers bobbles and glows: deep blue monkshood, purple-blue delphinium, sky blue flax, yellow columbine, golden sunflower, pale lousewort, elephant's head, shooting star, red skyrocket, Indian paintbrush.

The young plant is out there somewhere, having already flowered, waving slightly in the breeze, exuding its sweet scent. Things are looking pretty good. Pollinators are interested. Pests aren't bad. The soil is excellent.

And the neighbors seem friendly.

ELEVEN

The Tower of Babel and the Tree of Life

BOTANISTS CAN SOUND PLEASANTLY archaic or irritatingly pompous. They may know the common name of a flower, a blackfoot daisy, but they say Asteraceae instead. They love the jangle of Latin and vie for the prize of most fluent. Is that a *Melampodium leucanthum?* A *Chrysanthemum leucanthemum?* A *Monoptilon bellioides?* A *Bellis perennis?*

No, not an *Eregeron divergens?*

The adults talk. The children are excluded.

"It's a *daisy*," the layperson mutters under her breath.

In traditional taxonomy, the daisy belongs in the kingdom Plantae, in the division Angiospermophyta, in the class Dicotyledoneae, in the order Asterales and in the family Asteraceae, in which there are over one thousand genera. These genera contain some nineteen thousand species. Each grouping (kingdom, division, class) is a taxon, the plural of which is taxa.

Taxonomists are people who group things together. Taxonomists belong in the kingdom Animalia, in the phylum Chordata, in the class Vertebrata, in the order Primata, in the family Homonoidea, in the genus *Homo*, in which there remains only one living species, *sapiens*.

The taxonomist scolds the layperson: There are too many daisies in the world. We need to be more specific.

For that matter, there are too many bluebells, the common name for various species in different genera: *Wahlenbergia saxicola* in New Zealand, *Phacelia whitlavia* in the United States, *Clitoria ternatea* in West Africa, *Campanula rotundifolia* in Scotland, and *Endymonion non-scriptus* in England.

In England alone, there are ten different cuckoo-flowers, all plants that blossom early when the cuckoo sings. There are too many cuckoo-flowers and too many silly common names, like touch-me-not or bloody cranesbill or open-arse or jack-in-the-pulpit or firewheel or scarlet bugler or witch hazel or lady's slipper or figswort or beardstongue or snakebroom.

The adults are trying to have a serious conversation, and the adults require a serious language.

THE FIRST KNOWN WRITTEN classification of plants was in Latin in the fourth century B.C. Almost two thousand years later, an English botanist produced the sec-

Oxeye Daisy

ond major classification, also in Latin. In the eighteenth century, one scientist confessed that when speaking to his father as a boy, he was *only* allowed to speak in Latin. In this way, he learned that tongue before his native Swedish.

"Help, Father! I'm drowning!"

"Filius, filius, linguá latiná dicte!"

Parenting like this involves a certain distance, some cold calculations, perhaps a rather controlling personality.

A similar style might have produced someone like Carl Linnaeus, born in 1707, also in Sweden. Linnaeus's father was a clergyman and an avid botanist, as were his uncle and grandfather. (His great-grandmother had also been a botanist, which caused her to be burned as a witch.) Linnaeus grew up to be a proud, vain, insecure man whose genius lay in organizing plants, which he grouped into twenty-four classes based on their reproductive parts.

"God creates," Linnaeus said, "Linnaeus arranges."

Arrogance may be necessary for cataloging the world. In the promotion of his work, Linnaeus also found it helpful to refer to himself in the third person. People took him seriously. Artificial, and sometimes strange, his system was still more convenient and more comprehensive than any of its time. In a matter of years, it became the standard.

Importantly, Linnaeus gave a binomial, or two-part name, to each species. This is the method we still use. The first part, or capitalized word, is the genus, for example, *Mentha* for mint or *Vitis* for grapes. The second, uncapitalized word is often descriptive: *Mentha peperita* for peppermint or *Vitis vinifera* for a common wine grape.

Today scientists follow a set of rules, the International Code of Botanical Nomenclature (the Botanical Code), whenever they discover a new species. The species is first assigned its place in the ranking of kingdom, division, class, order, family, and genus. The discoverer then uses the binomial system to choose a name. A description of the plant in what we call Botanical Latin and in the botanist's own language is submitted to the appropriate journal, where editors and others determine if the species is really new and the name really unique.

Botanical Latin has simplified the ancient grammar, added words, and changed the meaning of words. It would sound Greek to a classical Latin scholar or to an early Roman. Still, as one linguist notes, a living dog is better than a dead lion.

Botanical Latin is a living dog.

Eregeron divergens. Monoptilon bellioides.

The words that first clank and clatter in your mouth will soon gather smoothness, like rocks being smoothed by the water of a stream, tumbling, turning, weighted with history, suddenly light.

Melampodium leucanthum. Bellis perennis.

You, too, can join the conversation.

You, too, will feel special.

CARL LINNAEUS CLASSIFIED FLOWERS a hundred years before the publication of the theory of evolution. Linnaeus used physical features to group plants. Most of his classifications were obsolete by the nineteenth century. What remains is his binomial naming system and the hierarchy of ranks.

Modern taxonomists try to group plants based on their descent, in terms of how and when they evolved through time. Down the line, in each rank, organisms share more and more similar features until, at the level of the species, they share enough reproductive features to breed together and produce a fertile generation. Down the line, in each rank, organisms share a closer and closer evolutionary relationship.

Taxonomists agree that their groupings should reflect evolution. They do not all agree on which features of a plant are primitive and which are more recent, on which are important for designating a rank and which are not. For this reason, biologists use a number of different taxonomic systems, which reflect the different personalities of their taxonomists.

For example, plants that produce legumes or beans as fruit can be divided into three groups, based on their different flower forms. A mimosa flower is circular, or radially symmetrical. A common pea flower is bilaterally symmetrical with five petals; the middle petal is larger, distinct, and protrudes from the bud. A honey locust flower is also bilateral with five petals; the middle petal is neither large nor distinct nor exterior to the bud. If you think these flower shapes are important, you will divide these three groups into three families. People will call you a splitter. If you think these shapes are not important, if you believe that the fruit or legume is what matters, you will lump all these plants into one family, with three subfamilies. People will call you a lumper.

You may use another approach to develop your taxonomy: cladistics. First you find a specific aspect of a plant that you can theorize as primitive or derived. The red flower of a rose and of a hedgehog cactus are produced by two chemically different pigments. Which pigment came first and which came from the other? Different cladograms, or pictures, will show different patterns or branches of evolution. With a complex question involving many factors, there may be hundreds or thousands or millions of possibilities. Computers do the math, helping you guess which alternative is most likely.

New tools are available to all taxonomists. We can now look into the DNA of a plant and gauge when, how, and from what it evolved. We examine cells. We examine genes. We examine chromosomes. We draw back from the microscope with some dismay.

Many of our current classifications are wrong. Every day, there is more bad news: The lotus is *not* related to the water lily. Lotuses belong with sycamores.

One little white-flowering mustard, *Arabidopsis thaliana*, is commonly used by laboratory scientists. The first plant to have its entire genome sequenced, it has been used in hundreds of studies and reports. We know an extraordinary amount about this plant. Unfortunately, when a botanist studied the twenty-five other species in its genus, twenty were found to have no evolutionary relationship. The plant would have to be renamed. Around the world, plant geneticists were appalled.

The sheer volume of new information about plants is becoming harder and harder to fit into the traditional system of ranks—kingdoms and phyla and classes and families and genera and species. Students used to memorize these ranks by reciting mnemonic phrases: King Philip Came Only For Gold and Silver. Then taxonomists began to add superorders, subfamilies, tribes, cohorts, phalanxes, subcohorts, and infraphalanxes, all of which require phrases too long to be mnemonic.

In the current system, new discoveries can have a domino effect. Changing the name of a plant's lineage may mean changing the names of any number of other plants. Changes in the Botanical Code can be complex and painful.

Simply naming a plant with the Botanical Code can be complex and painful.

Some biologists want a new system. They want to abandon the contorted, inconsistent grouping of organisms into ranks like kingdom, phylum, and class. Instead, a plant would be given a single proper name, which would indicate its evolutionary path. That name would retain the Linnaean sense of hierarchy, one group "nested" inside a larger group. Humans might be called *Sapiens Homo Homidae Primata Mammalia Vertebra Metazoa Eucaryota* or, for short, *Sapiens Homo*. If we learn something new (if we really do have extraterrestrial ancestors), changing that name will be easy.

No one would care about King Philip anymore.

"It's the greatest thing since sliced bread," says one botanist.

“It’s moronic,” says a colleague.
“It’s tempting,” says a third.
It’s another problem in communication.

WE WANT TO KNOW the daisy by naming it. Then the whole thing snowballs. We want to know the names of the daisy’s relatives and the names of the plants related to those plants and the names of the plants related to those plants, and we realize that we want to name the whole world.

The daisy is a tiny part of what botanists call the tree of life. We want to know the tree of life.

It’s more than finding a common language, more than having a conversation. It’s about how all the names begin to connect and interact, how the words form into something larger, like the cellular growth of a plant itself, budding into branches and leaves, an embrace of everything alive on this earth.

The tree of life. Commonly, we draw a picture in which the base of the tree starts with simple, single cells. These cells, or prokaryotes, have few internal structures. Some are adapted to live in difficult climates, like very hot pools or the harsh environment of a young planet. Above the prokaryotes, we have cells called eukaryotes, which are more complex, with a center or nucleus and other internal structures. Both these single-celled life forms make up, by far, the largest and most unexplored part of the tree. There are millions and millions of species here that we have not yet discovered or even bothered to discover. This enormous "trunk" is often divided into two kingdoms: Monera and Protista.

Higher now, toward the top, single complex cells combine into multicellular eukaryotes. This is what we call the crown of the tree. We used to divide this crown into three branches or kingdoms: Plantae, Fungi, and Animalia.

Lately, research has been playing among these branches like a great wind.

The single branch Plantae is really three separate groups, three lineages that evolved from three different one-celled organisms. Green plants are green algae and all land plants. Red plants are red algae. Brown plants are brown algae, diatoms, and organisms that look like plants but do not use photosynthesis.

Fungi, including yeasts and mushrooms, are a fourth branch. A mushroom may sit around and grow like a daisy. But mushrooms and other fungi are evolutionarily more related to animals than to plants.

Animals (drumroll, please) are a fifth branch. From a medical standpoint, our evolutionary closeness to fungi means that fungal infections are difficult to treat because whatever is harmful to the infection is often also harmful to us. Like kissing cousins, we have too much in common.

Knowing who goes where and who is related to whom on the tree of life is useful. Our response to a disease changes if we know it is bacterial and not fungal. In other cases, if a plant is helpful to us in some way, its relatives may also be helpful. When researchers found that the Pacific yew tree produced taxol, a cancer-fighting drug, the yew tree soon became endangered through overharvesting. Quickly we looked for related species that also produce taxol. We even found a species of fungus, living on the yew tree, that makes taxol as well.

Yew tree, fungus, and researcher: We are more closely related than we like to think. We huddle together on our part of the tree, surrounded by bacteria, outnumbered by microbes.

IN THE TREE OF LIFE, the human twig is small. Our kingdom, Animalia, is minuscule. Even so, we have a singular power. We're the mouse that roared. We roar out names. We dream about names. We praise those gods who gave us the names of all the animals and all the plants and all the living things on earth. We know that naming is magical. We know that naming is ownership.

A rose by any other name would smell as sweet. But that's just the opinion of one man. Maybe it would not smell as sweet. Maybe it would smell differently. Maybe a name makes all the difference in the world.

It's a daisy. It's pretty. Its center is the color of egg yolk. It has milky white petals. We pick those petals, one by one, and we whisper, "He loves me. He loves me not." We make a wreath to put on our heads.

We want to name the daisy.

It's *Eregeron divergens*. No, it's *Bellis perennis*. No, it's *Chrysanthemum leucanthemum*.

With some ceremony, we put the daisy in its rightful place on the tree of life.

TWELVE

Flowers and Dinosaurs

IT'S A BILLION YEARS AGO.

You are in the water and everything is easy and everything is right there, in the water, floating by. You release your male reproductive cells and they swim away. You release your female reproductive cells and they swim away. You leave the ocean. You come to a freshwater lake. It's nice here, too. You are as happy as the clam that hasn't evolved yet.

You're not vain. You don't think of yourself as the Eve of green plants. You're the size of a pinhead, one cell thick. You have this problem. The shoreline of the lake keeps changing. The water leaves and you dry up.

You adapt. You protect yourself from the sun. When the water rises, you send out sperm. When the water recedes, you sit there adapting. It's a new you. A new look. You no longer think of yourself as a green alga.

You like the word *moss*.

It's 500 million years ago.

You have no leaves. You have no stems. You have no roots. You have trouble transporting minerals and water. You decide to become a fern. Someday you will be a houseplant. Human slaves will mist you everyday. You will live in a big house with a view of the ocean.

You're a little vain now. You're intrigued by packaging. You'd like to protect your embryo. You call it a seed. You'd like to put your sperm cells in a container. You call it pollen.

It's 200 million years ago.

You have discovered aerodynamics. The wind takes your sperm to another plant, and by now you are a pine tree, a member of the biggest gang in town. You roll across the land undulating in huge forests. You are the most successful form of vegetation on earth. If you could find one your size, you'd wear a red and black jacket emblazoned with your team name: The Gymnosperms.

You are a gymnosperm. You don't have flowers or fruit. You are resting on your laurels, which also haven't evolved yet. You feel immensely proud.

You want to share your feelings. You look around the forest. You notice the dinosaurs.

Two hundred million years ago, at the beginning of the Jurassic period, gymnosperms and reptiles dominate plant and animal life on land. The group of reptiles known as dinosaurs has been around for a long time. In the next 60 million years, they will get bigger and bigger. Eventually some will be seventy-five feet long and weigh seventy tons. They roam in herds and walk on pillar-like legs. Their movements shake the ground. Under a hot sun, they browse the tops of coniferous trees, cycads, ginkgoes, and seed ferns, drawing leafy branches through their rakelike teeth and processing food slowly in their stomachs. At this time, most of the continents still huddle together in one big land mass. A few bees and other insects flit through the air. A few ratlike mammals scurry in the dirt.

The writer Loren Eiseley has described the scene: "Inland the monotonous green of the pine and spruce forests with their primitive cone flowers stretched everywhere. No grass hindered the fall of the naked seeds to earth. Great sequoias towered to the skies. The world of that time has a certain appeal but it is a giant's world, a world moving slowly like the reptiles who stalked magnificently among the boles of the trees."

In this monotonous green world, real flowers are starting to evolve, perhaps from seed ferns (a group of plants extinct today), perhaps from the shrubby bennettites (also extinct), or perhaps from plants whose descendants include members in the genus Ephedra.

Meanwhile the huge land mass is breaking apart. India rafts north. North America drifts west. By 140 million years ago, toward the end of the Jurassic period, the ovules of some plants may be beginning to develop fleshy carpels that enclose and protect their previously naked seeds. The seeds of these plants spread and catch a ride on the drifting continents.

Across the seafloor, the passage of North America is triggering tremendous volcanic activity. Mountain ranges rise. Africa and South America separate. The shallow seas of Europe start to drain away.

Too much is happening. There is too much time.

This is why we like fossils. A fossil is limited to the life of a single plant or animal at a specific moment. I can see myself in a fossil: smothered in silt, water squeezed from every cell, compressed, hardening, waiting for a scientist to find me. I can understand a single fate.

One of the earliest flower fossils is 120 million years old, from Koonwarra, Australia. Scientists first thought this plant was a no-count fern frond. Then someone noticed flower clusters the size of a period. The tiny florets are all female, without sepals, petals, or stamens. The single carpel has no style and is protected by reduced leaves. The entire fossil is 1½ inch long and looks like a small, black pepper plant. Its male counterpart is presumably somewhere, buried in rock. Most likely, the Koonwarra flower was pollinated by the wind. Perhaps small insects were also involved.

At this time, during the early Cretaceous period, beetles are already pollinating cycads with their palmlike leaves. Other gymnosperms may be using pollinators as well. Certainly, bees and flies have been around as long as the dinosaurs.

Unaccountably, the dinosaurs are starting to shrink in size. Since smaller animals lose heat more rapidly, dinosaurs now need a higher metabolism. Herbivores begin to chew more efficiently. Nostrils get bigger and breathing also becomes more efficient. The breathing passages separate from the mouth cavity. Now dinosaurs can chew and breathe at the same time, quickly digesting food and transforming it into energy. The ratio of brain size to body weight increases. Behavior becomes more flexible.

By the late Cretaceous, dinosaurs have diversified to such an extent that there are more kinds of them, small to big, than at any other time.

Flowers are also diversifying. In Asia and North America, they are leaving fossil records that are 110 million years old. Some flowers have both male and female organs. Some have eight carpels, not just one. Some carpels have fused.

In New Jersey, 90 million years ago, a fire hardened the cell walls of hundreds of flowers trapped in muck. The muck turned to stone. These tiny blossoms are three-dimensional. There are plants here similar to magnolias with spirally arranged floral parts. There are the fossilized remains of weevils, suggesting that some flowers were pollinated in a typical "mess and soil" beetle style, with insects wandering over the plant, eating, mating, defecating, picking up pollen, and carrying pollen, just as beetles do today. There are flowers that packaged their pollen in a mass, something flowers do when they have a pollinator they trust. There are flowers related to modern-day rhododendrons, hydrangeas, carnations, azaleas, pitcher plants, and oaks. There are flowers that offered resin as a reward.

This is a bouquet from the Late Cretaceous.

Everywhere now, during the Late Cretaceous, on the drifting continents, flowering plants have established themselves as weedy herbs and shrubs, able to colonize open or disturbed areas. Flowers are developing bilateral symmetry. Petals are fusing into shapes that small creatures can crawl into. Insects are responding and changing. They discover a nectar gland. They flit from one sugary drink to the next. A moth swoops lazily past a dinosaur's nose. It lands on a pleasing, sweet-smelling blossom.

Dinosaurs, oddly, may have helped make this possible.

In the Jurassic period, the dominant herbivores were giant creatures that ate the tops of gymnosperms. Usually these large plants could survive this assault, while their seedlings and saplings matured below.

But later, during the Cretaceous period, the dominant herbivore became smaller and shorter, with a large, muscular head and flat, grinding teeth for chewing plant tissue. These herbivores may have eaten the lower, younger trees, perhaps before they could mature and seed. These dinosaurs interacted with gymnosperms in a very different way.

And gymnosperms, throughout the Cretaceous period, began to decline. Suddenly the fast-growing herb and shrub, the little angiosperm, the intrepid colonizer, had the advantage. As the great forests of gymnosperms died out, even more habitat opened up, new places for flowers to grow and evolve.

In Loren Eiseley's famous essay, "How Flowers Changed the World," written in 1972, flowering plants provided small mammals a new high-energy food source: nectar, pollen, seeds, and fruit. In concentrated form, these foods would allow mammals to expand and flourish. After the tiny flowers of grass appeared, there would be plains of grazers, all mammals now, and the quick rush of a furry predator. Millions of years later, at the transitional edge between plain and wood, a particularly curious mammal would stand upright, and stare, holding a stick.

Eiseley ended his essay with this memorable sentence: "The weight of a petal had changed the face of the world and made it ours."

We remember this sentence when we look at a weedy mustard, fading yellow, or a ragged poppy dusty on the roadside. Flowers may have prepared the way for you and me.

Dinosaurs may have prepared the way for flowers.

At least once in our lives, we have imagined ourselves in the world of dinosaurs. Specifically, we are hiding in the foliage, 65 million years ago, while *Tyrannosaurus rex* roars nearby.

Here she comes, closer and closer.

T. rex is a dinosaur with a street name. *T. rex* moves like a madwoman, jerky, purposeful, looking over her shoulder. She is forty-five feet long and nineteen feet high, and she weighs four tons. Mostly we imagine her jaws and teeth. We have atavistic memories, bolstered by TV. We remember what it was like to be eaten alive.

If we looked around, away from the teeth, we would see a familiar landscape. There are conifers like bald cypress, redwood, and cedar. There are also sycamores, laurels, tulip trees, and magnolia trees. We don't see grass yet, and we don't see anything like sunflowers. But we do see a lot of flowering plants. The Age of Gymnosperms is almost over.

Tyrannosaurus rex stomps out of sight, muttering in the tulip trees. She doesn't know that more than a third of all animal and plant species on earth and two-thirds on land are about to go extinct, that she lives at what paleontologists call the KT Boundary, the thin border of time between the end of the Cretaceous and the beginning of the Tertiary period. She doesn't know that when paleontologists talk about "what crossed over the KT," what survived, they do not mention her name.

NO ONE REALLY KNOWS what happened.

For many years, volcanoes had been erupting, violently and continuously. These eruptions might have injected poisons into the air. They might have caused global cooling. Meanwhile, shallow seas were withdrawing from the continents. Among the dinosaurs, there might have been disease and a dwindling gene pool. Dinosaurs might even have been bothered by those small, pesky, egg-eating mammals.

Certainly they were bothered, at the end of the Cretaceous, when an asteroid hit the earth. In the Yucatán Peninsula, the impact crater is 120 miles wide. The crash sprayed debris all over North America and threw comet-enriched material or iridium around the world. Large areas were devastated by fire. For months, everywhere, dust and ash probably obscured the sun, while chemicals caused global acidic rain.

In this scenario, dinosaurs died quickly. Smaller reptiles and mammals, feeding on dead and decaying plants, "crossed over." Certain seeds "crossed over."

In one site in North Dakota, 80 percent of plant species disappeared. This number is based on the fossilized plant parts seen below and above the KT Boundary, which is marked by iridium and shocked mineral grains, remnants of the asteroid crash. Just above the boundary, fern spores increase, and ferns may have been a post-impact recovery plant. For a short while, fern prairies may have dominated the landscape.

In another site, in the Russian Far East, only one angiosperm crossed the boundary.

Dinosaurs were gone. Flowers were devastated.

According to some people, you can set your watch by it. After a mass extinction, evolution speeds up. Everyone diversifies. Everyone radiates. Mass extinctions are usually caused by violent changes in habitat and climate. These same changes often mean that the populations of surviving species become isolated from each other. New species evolve. The world fills up again.

The next period of time, the Tertiary period, or Age of Mammals, is also the Age of Flowers. The surviving angiosperms set the standard. New, evolving flowers raise the bar. In an early pea blossom, we see wings and a keel. Members of the philodendron family have spathes and chambers for catching insects. The tubes and spurs of flowers lengthen to accommodate new kinds of insects and birds and bats. Suddenly there are butterflies!

Loren Eiseley writes, "Impressive as the slow-moving, dim-brained dinosaurs had been, it is doubtful if their age had supported anything like the diversity of life that now rioted across the planet or flashed in and out among the trees."

SCIENTISTS HAVE BEEN searching for the oldest living flower. They compared the mutation of genes in the chloroplasts of hundreds of plant species. When a computer program shuffled those genes into rough time order, an obscure shrub, *Amborella,* fell to the bottom.

Amborella is a living fossil, the closest relative of the first flowering plant. It has small, creamy flowers and red fruit and is found only on a single island in the South Pacific. Some botanists think that *Amborella* looks much like the original model, the first flower that had all the parts of a flower.

The next-oldest living blossom may be the water lily. Next comes the star-anise, and then the magnolia.

These flowers, and every flower in the world, are the descendants of those able to cross the KT Boundary. In turn, they trace back millions of years ago to that first green plant, the size of a pinhead, one cell thick, in a freshwater lake.

You are an angiosperm. You survived the Big Crash that you don't want to talk about. You discovered that there was more to life. You turned into an orchid. Now you amaze even yourself. You drip perfume. You pretend to be a wasp. You have secret corridors. You delight the bee.

Sometimes, dimly, you remember the dinosaurs.

THIRTEEN

The Seventh Extinction

"EIGHTEEN DEAD."

"Twenty-eight dead."

"*Thirty-two* dead."

Day by day in late July 1999, my eleven-year-old son made these announcements, as I prepared to leave for the Sixteenth International Botanical Congress in St. Louis, Missouri. In a period of two weeks, a heat wave in the Midwest would kill 271 people. My son was counting the death toll in St. Louis alone.

Heat exhaustion causes fatigue, dizziness, nausea, headache, and cramps. The skin looks pale and clammy. Breathing becomes shallow. Pulse is rapid. In high humidity, the body is less able to cool itself through the evaporation of sweat. Clammy, sweaty skin turns hot and dry. Heat exhaustion is now heatstroke. The brain shuts down. It's out of control.

That July, the sick, old, and young in St. Louis, Missouri, were dying of heatstroke. Volunteers brought swamp coolers and air conditioners to the poorer parts of town. Some people refused help. They drew the curtains and locked the doors. Some people had air conditioners but would not use them. Some people never saw a volunteer. One woman woke every two hours to bathe her grandmother with a cold sponge. Just before dawn, the younger woman rose from her bed and went to the couch. Grandmother was dead.

The Sixteenth International Botanical Congress took place in downtown St. Louis at a large convention center. Four thousand scientists from one hundred countries met to talk about plants. They gave more than 1,500 presentations at 220 symposia in rooms so cold I wore a light sweater. The congress is held every six years, not often in the United States, not since 1969. It is a mega-event, a botanical mecca.

This year, it was a kind of dirge.

The president of the congress began by predicting that, if trends continue, one-third to two-thirds of all plant and animal species will be lost during the second half of the twenty-first century. A natural rate of extinction is about one species per million per year. The rate is now one thousand times that and will rise to as much as ten thousand times the natural rate.

So far, the earth has experienced six major mass extinctions, beginning with the Cambrian extinction over 500 million years ago. In 2050, my son will be sixty-three years old. He will see the beginning of the seventh extinction.

One could also argue that I am seeing the beginning, and he will see the end.

The sixth extinction occurred 65 million years ago, when the dinosaurs disappeared, along with two-thirds of other species on land. That disappearance is still a bit of a mystery.

The seventh extinction won't be a mystery at all. Our children can take careful notes.

Most of these losses will be in the tropical rain forests, an ecosystem we are losing so rapidly that in fifty years we can expect to have 5 percent of what we have now. I have been told over and over, often very cleverly, how many acres of rain forest are being cut down every minute of the day, how many acres for every breath I take, for every heartbeat.

I can never seem to remember that number.

The president of the International Botanical Congress outlined a seven-point plan that would slow the current rate of extinction. The plan involved money, organization, and research, nothing that isn't possible or reasonable. Throughout the congress, more plans would be revealed, all requiring money, organization, and research. In smokeless rooms, over conference tables, men and women were hatching plots to save the world. Among the elite, back-room deals were being made.

At least, I hoped so.

I sat in a lecture hall listening to a woman pinpoint how we have made a mess of things. Human beings have transformed 50 percent of the land surface of the planet.

We have doubled the amount of nitrogen in the environment and increased the amount of heat-trapping gases in the air. Scientists no longer argue whether global warming is real. Every year is the hottest on record. Every summer has a deadly heat wave.

The oceans are in serious trouble. Some fifty dead zones, areas with little or no oxygen, have appeared in our coastal waters. The largest in the Western Hemisphere is in the Gulf of Mexico and is caused by nitrogen and phosphorus flowing down the Mississippi River. Shorelines are eroding. Toxic algal blooms have increased. Over 60 percent of coral reefs, which sustain one-quarter of marine wildlife species, are threatened. Much of the damage is unseen and underappreciated. Commercial trawlers literally scrape up the seafloor.

What does it mean to clear-cut the ocean?

That July, newspapers claimed that the human population on earth had just reached six billion. Less than forty years ago, we were half that. In fifty years, we will be double that. I am one among six billion. My son will be one among twelve billion. We are clearly in overstock.

This is where it gets tricky. Look at my eleven-year-old son. He is adorable. He will be adorable when he is sixty-three years old. Not a single one of us is less valuable because there are so many of us.

In another room at the botanical congress, a man talked about alien invasions. Plants and animals and fungi travel around the world bringing disease and

Skyrocket with Broad Tailed Hummingbird

disharmony. Invading species are a major reason why we are losing species. Humans are the co-conspirator. Snakes ride on planes to islands that never had snakes. Viruses hop aboard the luggage. Sometimes, we knowingly introduce exotics.

Islands are particularly susceptible. Hawaii is the extinction capital of the world. This state contains one-third of the nation's at-risk plants. Half its species of wild birds are gone. Virtual islands—small, wild places surrounded by human development—reflect our larger problem of habitat fragmentation. We are creating islands everywhere.

Scientists talk about weedy species, like *Homo sapiens*, and the future of the planet as a planet of weeds. Hawaii's honeycreepers will be gone. The sparrow will remain. The water lily will be gone. The dandelion will remain.

Any questions?

About alien invasions?

Global warming?

Extinction?

A reporter stood up and addressed the scientist. "With everything you know, with everything you have told us, do you have any hope?"

The audience waited. We watched the scientist's face. We watched him blink and move his mouth. We watched him look down. We watched him look up. The moment passed when he could have said yes.

"That's an unfair question," the scientist said.

We have come to this: "Do you have hope?" is an unfair question.

WE KNOW OF over 250,000 species of flowering plants. There are many more we have not discovered. We think that 25 percent of green plants will go extinct in the next fifty years. One researcher estimates that a plant species disappears somewhere every week. In the United States, one in three plants is at risk. Many of these extinctions could be prevented.

But we don't have much hope of that.

When we estimate what will happen to flowers, we may be underestimating what is happening to pollinators, which also show a worldwide decline. The extinction of one species can cause a cascade that affects others. The male euglossine bee visits certain orchids to collect perfume to use in mating. The female euglossine bee has a long-distance trapline, in which she pollinates woody plants scattered in the forest. These plants are threatened by lumbering, grazing, and development. The bees are threatened for the same reasons, as are the orchids. The success of each species is linked to the success of each species.

We didn't kill off the passenger pigeon because we shot the last pigeon. We killed so many of them that they could no longer function as a group. As it turned out, this was the only way they could function.

Most flowers are more adaptable than passenger pigeons. At least, we hope so.

Every day, at the International Botanical Congress, I read about the heat wave in the newspaper. In Chicago, a fourteen-year-old boy lay in bed. He was very ill. To make matters worse, the utility company had turned off the electricity in his mother's apartment because she hadn't paid the bill. This may have been a misunderstanding between her and the landlord and the company. She had, after all, only recently moved into the apartment. The article said that the utility company was very, very sorry because after they turned off the electricity, after the temperature rose, in the middle of the heat wave, the mother could no longer cool down the sick boy.

I imagine this woman rushing from her apartment, trying to find help, full of rage and disbelief. This can't be happening!

The boy died while she was gone.

I cannot see the mother's face. But I can see the boy, waiting, lying on his bed, his skin hot. He knows he is dying. He is too sick to care much. But he knows.

FOURTEEN

What We Don't Know

WE GET UP EVERY DAY, surrounded by mystery and marvel, enthused by all the things we do not know. Life on earth has had four billion years to get this far. We woke up this morning to try and figure it out.

I am meeting Rob Raguso at the Arizona Desert Museum in Tucson to watch the hawkmoths as they appear at twilight among mounds of sacred datura, also called jimsonweed, thorn apple, and moonflower. The sacred datura has large, trumpet-shaped, silky blossoms, creamy white and pale lavender. When eaten, the flowers can cause hallucinations, blindness, and death. Their beauty, as in a myth, has two faces.

I have grown up with this gorgeous, sinister flower, and I never take it for granted. Whenever I see it, I want to gasp.

In the museum's "moth garden" are also patches of purple verbena, yellow sundrops, rock trumpets, pink

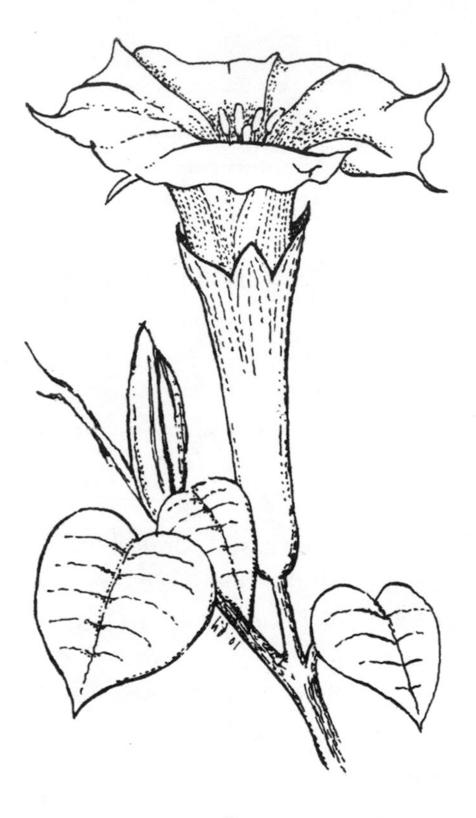

Datura

four-o'clocks, and white evening primrose. These primroses are delicate. The four heart-shaped petals seem tissue-thin, veined in a slightly darker color, like the veined skin of the very old or very young. The flowers look blown here, on their way to somewhere else. They look uncertain, as though they might blow away again.

In fact, they are busy, pumping out scent, preparing for the hawkmoth.

Hawkmoths are a group of species that appear worldwide. Coiled under the moth's head is a strawlike proboscis for sucking nectar. Their stout bodies are equipped with large, stiff, powerful wings boldly patterned. Hawkmoths have good eyesight in dim lighting. They can fly far, and they can fly fast. They can regulate their body temperature.

The hawkmoths in this desert, the white-lined sphinx moths, have four brown wings colored in bands of pink and white. Habitually, the larvae rise up like tiny sphinxes, daring you to interfere with their happy lives of eating everything in sight. These caterpillars are lime green with a yellow head, side rows of pale spots bordered by black lines, and a bright yellow or orange rear horn. They are strikingly beautiful. And they seem to know it.

As a child, Rob Raguso collected moths and butterflies. As an undergraduate at Yale, he learned about flowers as butterfly food. By the time he began his graduate work in the biology of floral scent, he was looking more at the food than at the butterflies, hoping someday to look at everything, to start with molecules released from a petal and end up in the sky pumping wings.

Rob is happy to talk about the relatively obscure wildflower Brewer's clarkia, the subject of his doctoral dissertation, completed in 1995. In the evening primrose family, *Clarkia breweri* is hot pink, with four petals that divide into a central lobe and two side lobes. The little flower looks gay and excitable. It grows fast and can be used easily for genetic studies. It is the only known scented species in a genus of over forty species.

"In the *Clarkia* genus," Rob says fondly, "we had a group of plants that were ancestrally bee-pollinated, that offered pollen to specialist bees, and that lacked fragrance. In time, *Clarkia breweri* evolved a long nectar tube, switched to nectar, and added scent. How did that happen?"

Rob spent a year learning how to collect and analyze odor molecules. He found that Brewer's clarkia has a relatively simple scent. The flower makes two kinds of chemical compounds: terpenoids, common in citrus and mint, and benzenoids, characteristic of clove and cinnamon. Trace amounts of terpenoids are also in the flower's nearest relative, *C. concinna*. Brewer's clarkia had amplified these compounds and added another group.

Next, Rob and others would determine which parts of the flower produced which odors, which enzymes were involved, and which genes. These are all things we did not know before. By adding scent, the bright pink flower attracted new pollinators, such as night-flying hawkmoths. By making a larger flower, having lots of nectar, and staying open in the day, Brewer's clarkia also began to attract hummingbirds.

But that's another story.

Rob was interested in whatever the hawkmoths were interested in. What, for example, did they smell? He learned how to record the responses of hawkmoth antennae and found that they smell everything.

This was something else we didn't know before. Rob thought it was wonderful news.

At the Tucson Desert Museum, Rob and I are joined briefly by his wife and three-month-old son. Rob talks to the baby as all parents do, as though the infant understood every word and might well answer in a complete sentence: "Yes, I *do* want to be changed right now." "No, I am not tired, although this is what you would like to believe. I *am* bothered by the setting sun in my eyes."

The setting sun drops behind the postcard purple, jagged mountains, and the world turns suddenly blurry. Like airplane pilots on a schedule, the hawkmoths appear, moving invisibly against a lattice of greenish gray leaves and ghostly flowers.

"Look, look," Rob says.

I try and see only the aftermath of wings, a ripple in the space-time continuum.

"Look, look," Rob urges.

I cannot see the hawkmoths, but I can smell the flowers and something more, a powdery sweetness I associate with my grandmother's makeup, her voice out of tune in the Methodist church as we sang together, "I Walk through the Garden Alone" and "Glory Be to the Power and to the Son aaand to the Ho-leee Ghost." All the cues of memory blend together in a complex system: the paisley of her dress, the polished wood of the pew, the sound of music.

Rob pulls a hawkmoth out of the air and holds it in his hand. I almost clap. That's a magic trick. The sphinx moth is the insect version of the hummingbird, whirring its wings, hovering to drink from a corolla tube. Briefly, the moth struggles.

"Look how muscular he is," Rob says admiringly. "This is a strong guy!"

The moth as action hero.

The air is suffused with scent, with sex, with food, with memory. "What is your question?" I ask Rob.

He pauses to let our hero escape.

"How does a moth experience a flower?"

SOME 200,000 YEARS AGO, human beings evolved to think creatively outside. Grass and sun and trees were the natural setting for thought. Carefully, we watched the other animals. Today, we experience delight when we are allowed to do this again. When we feel intelligent in a meadow, we feel right at home.

We are still evolving. Our best science now often happens indoors, in the laboratory, where we are surrounded by tools we invented but do not fully understand. In Rob's lab at the University of Arizona, gas chromatography analyzes the compounds produced by a flower. A computer program takes each compound's mass spectrum, its unique fingerprint, and compares it with the thousands of compounds we already know. There are thousands more we do not know.

Like a perfume maker, Rob has trained his nose to identify scent. He can equate a smell, a physical experience, with a molecule. Then he can equate that molecule with a mass spectrum.

It may be the closest any of us come to living like a hawkmoth.

Rob likes playing detective. When he smells a flower, he asks himself, What's in there? Why does it smell like grape or chocolate when it doesn't have any of those molecules? What is making this flower shimmer? What odors are masking other odors? What odors are synergizing to make a new odor? What does this odor mean to a honeybee?

One question leads to another and to six thousand more.

How does a moth experience a flower?

How does a flower experience a moth?

There are so many reasons to get out of bed.

Since hawkmoths respond to a range of odors, flowers can switch to hawkmoth pollination fairly easily. Flowers do not have to make a specific class of chemicals. They can simply smell good. Some plants amplify the emissions from their sepals and leaves. Some alter their defense compounds. Some use existing nectaries and anthers. Some evolve a new nectary.

Now Rob wonders what would happen if he compared three different plant families, all with night-blooming, moth-pollinated flowers: the evening primroses, the potato/tomatoes, and the four-o'clocks. How have these groups changed their strategies through time, losing and gaining fragrance as they evolved in different climates and soils? Is fragrance lost when self-pollination evolves? Can fragrance return? Are there patterns within each family?

The evening primrose family contains over 650 species. One of them is Brewer's clarkia. One is the flimsy-looking evening primrose I am watching now being pollinated in the Tucson Desert Museum. One is the tiny white enchanter's nightshade, which the Greek witch Circe used to change men into pigs.

We don't know much about these plants. Mostly we study crop plants for which our questions are quite practical. The four-o'clocks are still a mystery. The sacred datura is still a mystery. The white enchanter's nightshade is still a mystery.

How did she change those men into pigs?

AT THE INTERNATIONAL BOTANICAL CONGRESS, in a cavernous convention center, where five thousand scientists gather for seven days to talk about plants, I feel lost. I feel lonely.

Suddenly I see Rob. I have been at the conference for one hour, and here is a familiar face. I almost clap. That's a magic trick.

These big conferences have a hierarchy. Celebrity scientists give speeches at the plenary lectures. In the symposia, five to six scientists read and present research that is often already published. In the poster exhibition, usually the younger scientists and graduate students describe newer, unpublished work. They do this, unpretentiously, on a sheet of poster board.

At this congress, more than a thousand posters have been hung on six-foot-high panels that form rows through which people can walk and browse. Rob has a poster on evening primroses and hawkmoth pollination. He invites me to see him on Wednesday morning, 9–10 A.M., when the authors of even-numbered posters stand in front of their work and explain it to people passing by. Shortly after, the shift changes, for the authors of odd-numbered posters.

The exhibition hall is huge, with the ambiance of an airplane hanger. I enter the hall and am immediately beguiled: The place hums. The air hums and thrums. I have entered some kind of hive, a honeycomb of panels and posters, where people murmur and talk about flowers. The hive is busy. The hive is excited. So much work is getting done.

It's not all pure science. I roam, gathering phrases: "Relocation." "A very practical field." "He's difficult to work with." "My thesis committee." "A starting salary."

People are getting into new schools and finding new jobs and asking about mentors and fitting their lives into the life of the hive, of knowledge built on top of knowledge. I almost expect them to touch heads, smelling each other, exchanging news.

Rob's poster attracts a stream of men and women interested in primroses or in hawkmoths. Rob Raguso vibrates with enthusiasm. His dark eyes gleam. He has just gotten a new job at a university where he will teach some but where he will mostly do research, believing that all questions add up to something larger, believing that this is how he is meant to be in the world.

A woman stops and reads his poster. She, too, is interested in floral evolution. She, too, has worked with floral scent. She and Rob begin to dance.

FIFTEEN

Alchemy of a Blue Rose

PLANT GENETICISTS, WORKING for commercial flower companies, dream of creating a blue rose.

Why not a yellow smiley face on each petal? Why not red dots on a blue background with a yellow smiley face? . . .

Or would that be too busy?

THE TRUTH IS THAT we already have blue roses. I own a couch covered with them. I can walk into any department store and find some version of a blue rose, as well as many other things that human beings have shaped and colored. I enjoy buying these things.

But I have so many things.

To be honest, I don't even like double roses, flowers with an excess of petals crowding the carpel. Like most highly cultivated flowers, double roses are a mistake.

Some gene sent the wrong message to that part of the rose meant to develop into a stamen. Instead, the potential stamen acquired pigment and turned into a petal. Along its side, you can still find the flap that would have been the anther meant to hold pollen.

Obviously, such a mutation is inefficient in producing offspring and should normally die out. But for hundreds of years, gardeners have encouraged this kind of change and have cross-bred roses to produce a stunning array of extra petals, new colors, and prize-winning shapes.

Stamens can rather easily become petals. Evolution used the same idea for the original rose, in which the petals probably developed out of the stamens that grew next to the sepals. In that case, the mutation was beneficial. A few colorful petals seemed to draw more pollinators. (In other flowers, petals more clearly derived from the sepals themselves.)

In roses, we happily exchanged reproduction for decoration. What we lost was scent. Most roses don't smell sweet anymore. As it turns out, it is hard to put scent back into a flower through cross-breeding. Apparently, in the world of flowers, pollinators, pheromones, and odor plumes, smelling good is a more complex process than looking good.

Most flowers in private gardens and public landscapes have been cross-bred to look good, to get bigger, grow taller, bloom longer, stand up straight, think positive, and smile, smile, smile!

Many colors in a petunia or an impatiens would never be found in a forest or meadow. Some colors have been developed specifically, in the words of one plant breeder, to "go well with brick or nonwhite siding." They are the product of human thought and human labor. We hand-pollinate a promising plant with the pollen of another promising plant, perhaps from another closely related species, hoping for a hybrid with the right qualities, a more marketable penstemon or a yellow impatiens.

Impatiens are an extremely successful bedding plant. But they don't come, yet, in the color yellow. A single seed of such a flower, a commercial yellow impatiens, would be worth a lot of money. Americans alone spend billions of dollars each year on flowering plants and shrubs. The majority of these are hybrids. Annually, about one thousand new hybrid plants are introduced into the commercial flower market.

Many flowers in private gardens and public landscapes are aliens. If I visit a city in a hot, dry climate, I will see plants from hot, dry climates around the world. Flowers from Brazil live in Los Angeles. Flowers from China live in Ann Arbor. These flowers, too, have usually been cross-bred and domesticated for use.

Aliens can become too successful. There are dangers.

Still, it's really rather wonderful. Bougainvillea and bird-of-paradise fill my patio, where, for a nanosecond, I am a Hawaiian queen. I buy a sacred lotus for the miniature pond. When its petals unfurl, the Supreme God is revealed. I plant tiger lilies and a flamboyant hibiscus tree. It's a designer Eden. Geography trembles. Ecosystems mix. Imagination is as real as a plant. This patio is rich and trembling, with imagination, with unnatural connections, with a certain bravado.

A BLUE ROSE, OF COURSE, is not a simple matter of cross-breeding. A gene in a petunia contains the code for the enzyme that creates the pigment delphinidin, responsible for the color blue in flowers like petunias, iris, violets, and morning glory. In 1991, a flower company cloned the gene and inserted it into a rose. Not much happened. Possibly the delphinidin was being masked by other pigments in the rose. Also, delphinidin molecules may only be blue at a high pH (a low acidity), and most rose petals are too acidic. The company now hopes to find the genes that control pH in a petal or to crossbreed its variety with roses that are naturally less acidic.

Already, the company has used its cloned and patented blue gene to create a violet-colored carnation. A black carnation developed by the firm is also poised to enter the flower market. And the company has another carnation that can last a month in a vase on your dining room table.

As with any new hybrid, regulators in the United States and in Europe had to approve the violet carnation for sale. This was not a problem. Apparently the regulators did not believe that genetic material from a violet carnation could easily escape and get passed on to other plants in the environment. Genetically engineered carnations don't produce much pollen, which is buried deep in the flower. A cut carnation stops producing pollen. Furthermore, if (unlikely enough) a nearby weedy relative of the carnation were pollinated with a genetically violet cousin and if it produced fertile seeds—new, violet weeds—there was a sense of "So what?"

People don't worry much about violet carnations or blue roses. They feel differently about genetically engineered crops. Let's say that a crop is given a gene resistant to herbicides. Now we can spray that crop and kill only weeds. The fear is that the crop will hybridize with nearby plants to create a superweed, resistant to herbicides, or to pests, or to whatever has been engineered into the crop.

Or to something we didn't predict.

A gene from a common soil bacterium has been spliced into corn to create a crop resistant to the corn borer. Millions of acres in the United States have been planted with this corn, and the same gene is regularly spliced into potatoes and cotton. Only in the late 1990s did we realize that pollen from these crops is poisonous to monarch butterflies.

We are interfering with relationships we do not understand. This is not a news flash. We began interfering as soon as we picked up a rock and chipped it into a spear point. We were hell-bent on transforming the world. We haven't looked back.

This is who we are.

I don't want a blue rose in my garden. But I like the color blue. A small perennial herb called the dayflower grows in the mountains near my home. Its triangle of three petals is about an inch long, somewhat deeper than cerulean, lighter than indigo, more like ultramarine. Opening at dawn and wilting by noon, the flower rises delicately out of a tapered, folded leaf. Some people call the dayflower widow's tears, perhaps because of this tear-shaped leaf, or perhaps for some other reason. The dayflower does not grow in abundance. It seems rare. It appears suddenly, twinkling in the grass.

What did I feel the first time, and every time since, that I saw a dayflower, so singular and elegant, outside all that I know and am? The dayflower is mystery in a pretty shade of blue. The dayflower is the shape of the Other, the Beloved. If you are inclined, you could see God in this flower. You could feel transparent, clear as glass. You might even feel, dimly, what it will be like not to exist in your present form.

A blue rose is not the Other. A blue rose is an interesting artifact in a pretty shade of blue, perhaps the right color for that spot in the garden, yes, against the white wall, but not so good, unfortunately, with the patio brick—and clashing, really dreadfully, beside the bougainvillea.

WRITER AND CULTURAL CRITIC Jeremy Rifkin promotes the word *algeny,* which means "a change in the essence of a living thing." Algeny is analogous to the medieval idea of alchemy.

Alchemists in the Middle Ages believed that all chemical elements were transformable into other elements.

Nature was a continuum that we could ride, like an escalator. Furthermore, and of greater interest, all metals were on their way to becoming gold. This last transformation became a powerful symbol. Humans, too, could transform into spirit.

Jeremy Rifkin writes:

> The algenic arts are dedicated to the improvement of existing organisms and the design of wholly new ones with the intent of perfecting their performance. But algeny is much more. It is humanity's attempt to give metaphysical meaning to its emerging relationship with nature. Algeny is a way of thinking about nature and it is this new way of thinking that sets the course for the next great epoch in history.

The blue rose is part of the next great epoch.

With biotechnology, roses can be made to smell again. A commercial flower company has inserted into a rose the gene that encodes the enzyme used by citrus plants. The resulting rose will smell lemony. Eventually, we will program other fragrances into flowers. The blue rose can smell like cinnamon, like baking bread, like the talc-dusted skin of your first child.

Our ability to isolate genes, clone them, and put them into other plants has greatly sped up all kinds of research. Having sequenced the genome of the white-flowered *Arabidopsis thaliana,* scientists everywhere are popping genes in and out of this little mustard to see what happens. In a generation of seedlings, we can see the effect of a gene's absence, its presence, and its addition (or overexpression).

A particular gene, for example, with the nickname of ANT, controls the size of leaves and flowers. When ANT is inserted into the genome of a plant, the plant grows to produce bigger flowers and seeds. When ANT is removed from a genome, the resulting flowers and seeds are smaller.

The development of a flower is one of the things we understand least about plants. As we keep tinkering, however, we learn more and more, every day. We discover that one gene starts the response to a growth hormone. The mutation of another gene causes changes in the ovary. How about, now, that gene there?

Flowers are quickly giving up their secrets.

In the future, the crops we plant may flower at the times we choose, under the conditions we choose, in ways we choose. In our gardens, we will control the color of a flower, the shape of its petals, and the memories in its smell.

The blue rose, certainly, will do what we tell it to do.

I AM NOT ALWAYS SURE what I feel.

In Los Angeles, there is a garden center that I sometimes visit. Here, next to the freeway, flowers are crammed together row after row, one bright bloom after another: tulip trees, gardenias, fuchsias, hydrangeas, jasmine, wisteria, lilies, impatiens, vincas, zinnias, dahlias, verbena, daisies, hibiscus, and roses, roses, roses. The vast majority of these are hybrids. Many have signs attached, reminding me, "Asexual production of plants protected by the Plant Patent Act is prohibited."

Soon some of these plants will be genetically engineered.

I have stood in this place, surrounded by flowers, and I have been moved to tears. I have felt the excitement. So much beauty. So much bounty. It just went on and on, the beauty and the bounty, the alchemy and the algeny, all the magical arts. My heart beat faster. My chest felt hollow.

Flowers are on a fast-track continuum.

SIXTEEN

Phytoremediation

I AM SITTING NAKED in a hot spring. The water is a delicious 103 degrees Fahrenheit, mint smells strongly, cottonwoods and alders leaf above my head, yellow cliffs crumble above the trees, blue sky is above everything. Sliding deeper into the pool's warmth, I cushion my head against a rock and enter the drama of the bank: a tiny flower, an ant, another ant, a confrontation.

My friend next to me is also naked. Her pale legs shift and stir up mud so that a brown wash darkens her left breast. This hot spring is one of many in this canyon where, at the turn of the century, a sanitarium for consumptives was built around these heated pools. People wanted nature to cure them. They came for the sun and the air and the power of the land. Some were cured, and some were not.

After a few years, the sanitarium failed. The land was sold as a cattle ranch, and the cattle ranch failed, and a group of hippies bought this place in the 1970s with the dream of creating an international community, another kind of cure. The children of these hippies are still here. They walk about naked when they choose and take long baths in the hot water.

From this mint pool, a canyon runs northeast. My friend and I decide to walk under the brown and yellow cliffs, no higher than a two-story house. A small stream snakes over rounded rocks and soft sand. Barefoot, naked, we go slowly, from rock to rock. A juniper reaches out to catch my skin. Tall grass scuffles in the shadows between sun and stone. I feel, suddenly, alienated from this world.

We want nature to cure us.

My friend says, no, she would rather I did not write about her body, so I must write about mine. It is ordinary and I think about it in ordinary ways, the stomach too soft, the breasts nice. I see cellulite when I turn in a certain way. I am self-conscious, and I know this is odd: No one is watching me but me. My friend moves easily with her bare thighs. She lives in a place where naked is normal.

FOR A LONG TIME NOW, flowers have cured us in very direct ways. A quarter of our prescription drugs contain some part or synthesis of a flowering plant. At the same time, only 1 percent of plant species in the world have been studied for their medical use.

In folk medicine, the rosy periwinkle in Madagascar was prescribed for diabetes. When researchers began studying the flower, they found that extracts of the plant also reduced white blood cell counts and suppressed bone marrow activity. These experiments led to the isolation of two chemicals now used against childhood leukemia. With these drugs, a child's survival rate has increased from 10 to 95 percent.

For centuries, healers in Africa recommended a fruit called bitter kola for infections. In the 1990s, Nigerian scientists discovered that compounds of bitter kola may be effective against the Ebola virus, which causes a fatal disease characterized by severe bleeding. Ebola is a symbol of all the horrific diseases ahead of us, viruses that have mutated, epidemics that rise out of the jungle and the places we disturb. We have had no defense against the Ebola virus. Now, we may have the bitter kola.

On my walk with my friend, through a canyon in New Mexico, we stop before a ragged Emory oak, its gray-green leaves pointed, their edges sharp. All parts of all oaks have an antiseptic effect. Oak is the basic astringent, a wash for inflammations, a gargle for sore throats, a dressing for cuts.

All around me are plants that heal and connect to the human body. The yucca spiking above is a steroid. Mullein acts as a mild sedative. Mullein root increases the tone of the bladder. Juniper is used for cystitis. Yarrow clots blood.

My body is interwoven into the chemistry of juniper and yarrow. The tone of my bladder is related to mullein root.

How can we doubt our place in the natural world?

From every habitat, I hear a chorus of cures. In the American West, for menstrual cramps, I might take angelica, cornflower, cow parsnip, evening primrose, licorice, motherwort, pennyroyal, peony, poleo, raspberry, storksbill, or wormwood. For tonsillitis, I could try cachana, cranesbill, mallow, potentilla, red root, or sage. For a sunburn, I might turn to penstemon and prickly poppy. The juice of the prickly poppy was once used to treat a cloudy cornea. The poppy helps, as well, with inflammations of the prostate.

I stand in a canyon of crumbling yellow cliffs, embarrassed to be without my clothes, my soft stomach showing, my vanity showing, my prudery showing. Where else, besides my bed and bath, would I stand like this, exposed?

In the doctor's office. In the hospital, in illness and pain. To be cured there, I must also be naked. I must let myself be seen.

IN THE FIRST HALF of the twentieth century, the physician Edward Bach discovered in himself an unusual sensitivity to plants. He felt calmed or relieved near certain flowers. Others made him nauseous. Bach came to believe that the "liquid energy" of a flower steeped in spring water, warmed in sunlight, and mixed with brandy could cure the emotional problems at the root of human disease. He came up with thirty-eight flower remedies, mostly found within a few miles of his home. They are classed into seven groups for problems such as fear, uncertainty, "insufficient interest in present circumstances," "over-sensitivity to influences and ideas," despondency, and "over-care of the welfare of others."

The seven groups divide into finer categories. The monkeyflower is for fears you can name, whereas the catkins of aspen are for vaguer feelings of dread. Clematis restores those who live more in dreams than in reality. Honeysuckle restores those who live in the past. Wild chestnut is the remedy for the woman obsessed with a repetitive thought. Violet, impatiens, and heather are suggested for loneliness.

The Bach Flower Remedies are still sold today. They are based on the belief that our biochemical, cellular self is further fine-tuned by other, subtler energies, what the Chinese call ch'i and the Indians call prana, absorbed through the meridian and chakra systems. Flowers influence this energy flow. Flowers raise vibrations and open channels. Flowers act as a catalyst for change.

Over the years, people have added to Bach's work. The sunflower was not on the original list. Today, as a flower remedy, the essence of sunflower is recommended for people who suffer from arrogance, as well as for people who have low self-esteem.

The Bach Flower Remedies are easy to make fun of. They almost seem to make fun of themselves. But I do not want to make fun of them. At least, not too much.

I take all this as metaphor, and I take metaphor as the essence of how we think and live. I also believe that a sunflower can cure arrogance. I know for a fact that violets make me less lonely.

PHYTOREMEDIATION COMES from the word *phyto*, meaning plants, and *remediation,* the act of repairing or healing. Phytoremediation is a new field in science and a new business investment. Certain plants have the ability to take in and absorb toxic metals, which the stems and leaves of the plant hold safely in their cells and use in defense against insects or infection. These plants are now being used to clean up polluted soil.

In a Boston suburb, alpine pennycress drew up lead, zinc, and cadmium from a backyard where children were not allowed to play. Most plants cannot tolerate more than 500 parts per million dry weight of zinc. But pennycress stores up to 25,000 parts per million. At an abandoned zinc smelting plant, pennycress increased its rate of absorption of zinc and cadmium in the second and third year. The now-contaminated plants were then uprooted and safely destroyed.

Other flowering plants are being considered for a variety of jobs. Poplar trees have been used to remove chlorinated solvents in groundwater. Clover may remove petroleum. In India, aquatic plants deal with the chromium produced by tanneries. Some plants can defuse explosive compounds like TNT in the soil. Sacred datura takes up heavy metals like lead. Cabbage can reduce radioactive particles.

Sunflowers also absorb and store radioactive material. A company in New Jersey used sunflowers to decontaminate a uranium factory. In their hydroponic tanks, the roots of the sunflowers created a bio-filter for wastewater. In experiments in Chernobyl, sunflowers absorbed 95 percent of radioactive strontium in a pool near the leaky reactor. In 1996, the U.S. secretary of defense and the Ukrainian defense minister ceremoniously sprinkled sunflower seeds over a former missile silo.

In the United States, the sunflower remains an important economic crop, grown for its seeds and oil. Fields of sunflowers unfurl across the American Midwest like great blazing banners of yellow and orange.

The Incas of Peru used to worship the sunflower as a symbol of the sun and their sun god.

In gardens, right now, people grow sunflowers and fall down before them. These people are stunned, compelled to worship again.

WE MAY NEED to be cured by flowers.

We may need to strip naked and let the petals fall on our shoulders, down our bellies, against our thighs. We may need to lie naked in fields of wildflowers. We may need to walk naked through beauty. We may need to walk naked through color. We may need to walk naked through scent. We may need to walk naked through sex and death. We may need to feel beauty on our skin. We may need to walk the pollen path, among the flowers that are everywhere.

We can still smell our grandmother's garden. Our grandmother is still alive.

Selected Bibliography and Notes

Flower in the crannied wall,
I pluck you out of the crannies;
Hold you here, root and all, in my hand,
Little flower—but if I could understand
What you are, root and all, and all in all,
I should know what God is and man is.
Lord Alfred Tennyson

The Physics of Beauty

The information on Neanderthal burials was mainly taken from Arlette Leroi-Gourhan, "The Flowers Found in Shanidar IV, a Neanderthal Burial in Iraq," *Science* 190 (November 1975).

The quote by Annie Dillard is from her collection of essays *Teaching a Stone to Talk: Encounters and Expeditions* (New York: Harper Collins Books, 1982).

The quote from Aldo Leopold is from his *Sand County Almanac* (New York: Oxford University Press, 1949).

Frederick Turner, *Rebirth of Value: Meditations on Beauty, Ecology, Religion, and Education* (State University of New York Press, 1991), has a longer list of the themes or tendencies of the universe. These are also explored in Brian Swimme and

Thomas Berry, *The Universe Story: From the Primordial Flaring Forth to the Ecozoic Era* (San Francisco: Harper, 1992), as well as in other books and articles.

More about the giant arum can be found in textbooks and in Susan Milius, "The Science of Big, Weird Flowers," *Science News* 156 (September 11, 1999).

A college textbook I referred to frequently and recommend is Randy Moore et al., *Botany* (Wm. C. Brown, 1995). Their selection "Leonardo the Blockhead" summarizes the math behind a sunflower's seed spirals. This material is also known as the Fibonacci series.

Andy Coghlan, "Sensitive Flower," *New Scientist*, September 26, 1998, nicely summarizes recent work on how flowers "see," "smell," "touch," and "taste." Numerous others, such as Stephen Day, "The Sweet Smell of Death," *New Scientist,* September 7, 1996; Garry C. Whitelan and Paul E. Devlan, "Light Signaling in *Arabidopsis,*" *Plant Physiology Biochemistry* 36 (1998), issue 1–2; and Paul Simons, "The Secret Feelings of Plants," *New Scientist,* October 17, 1992, discuss these subjects in depth.

Dagmar von Helversen and Otto von Helversen, "Acoustic Guide in a Bat-Pollinated Flower," *Nature*, April 29, 1999, give more information on how bats and flowers use sonar signals.

The Blind Voyeur

The man who lost color sight as a result of brain damage is described in Oliver Sacks, *An Anthropologist on Mars* (New York: Alfred Knopf, 1995).

Moore et al., *Botany,* provided me with a good description of the visible spectrum and the function of pigments in a flower petal.

Deni Brown, *Alba: The Book of White Flowers* (Portland, OR: Timber Press,1989), is a comprehensive discussion of white flowers and includes passages on how and why white flowers look white.

Rob Nicholson, "The Blackest Flower in the World," *Natural History* 108 (May 1999), is my source for the information on the Oaxacan flower.

Moore et al., *Botany,* includes a passage called "Why Plants Are Not Black."

The information on bees comes from many sources. Naturally I consulted the seminal works of Karl von Frisch, including his *Bees: Their Vision, Chemical Senses, and Language* (Ithaca, N.Y.: Cornell University Press, 1971); and *The Dance Language and Orientation of Bees* (Belknap Press, Cambridge, MA 1967).

A translated version of Georgii A. Mazokhin-Porshnyakov, *Insect Vision* (Plenum Press, New York, NY 1969), was helpful for background and a sense of history.

An excellent book that I often referred to is Friedrich G. Barth, *Insects and Flowers: The Biology of a Partnership* (Princeton, N.J.: Princeton University Press, 1991).

Another important source for the behavior and physiology of pollinators is Michael Proctor, Peter Yeo, and Andrew Lack, *The Natural History of Pollination* (Portland, OR: Timber Press, 1996).

Lars Chittka was another primary source and a great help. Lars has pioneered much of the most recent research in insect vision, particularly in terms of what colors bees actually see. His most pertinent articles, related to this chapter, are Lars Chittka and Randolf Menzel, "The Evolutionary Adaptation of Flower Colours and the Insect Pollinators' Colour Vision," *Journal of Comparative Physiology A 171* (1992); Lars Chittka, Avi Shmida, Nikolaus Troje, and Randolf Menzel, "Ultraviolet as a Component of Flower Reflections and the

Colour Perception of Hymenoptera," *Vision Resolution* 34, no. 11, p. 1489–1508 (1994); Lars Chittka and Nickolas Waser, "Why Red Flowers Are Not Invisible to Bees," *Israel Journal of Plant Sciences* 45 (1997); Peter Kevan, Martin Giurfa, and Lars Chittka, "Why Are There So Many and So Few White Flowers?" *Trends in Plant Sciences* 1 (August 1996); Lars Chittka, "Bee Color Vision Is Optimal for Coding Flower Color, but Flower Colors Are Not Optimal for Being Coded: Why?" *Israel Journal of Plant Sciences* 45 (1997); and Lars Chittka and Nickolas Waser, "Bedazzled by Flowers," *Nature*, August 27, 1998.

An earlier version of this chapter contained a longer version of why white flowers look bee-green and why green leaves look gray: "White flowers that reflect UV are actually rare. Most human-white flowers, like this daisy, absorb UV. They do not look bee-white, since they are not reflecting back the bee's entire spectrum. They are reflecting back blue and green and, to a bee, look blue-green. To a bee, the green, serrated leaves of a daisy probably look gray. In bee vision, green foliage has a weak and uniform reflectance, which makes it dull or uncolored. For humans, leaves absorb relatively more light in the red range."

The idea that bees had color vision before the appearance of flowers also comes from Lars Chittka's research, described in Kathleen Spiessbach, "The Eyes of Bees," *Discover*, September 1996. An example of Chittka's style and humor can be found in his article about flower color coding in Chittka, "Bee Color Vision": "But how can we determine in what colors insects saw the world 200 myr (million years) ago? Since time machine projects habitually run into complications (e.g., Wells, 1885), it is now difficult to obtain funding for them and so evolutionary biologists resort to an alternative strategy called comparative phylogenetic analysis."

Nickolas Waser was also an important source for this and other chapters. A related article is Nickolas Waser, Elvia Melendrez-Ackerman, and Diane Campbell, "Hummingbird Behavior and Mechanism of Selection on Flower Color in *Ipomopsis*," *Ecology* 78, no. 8 (1998).

I should also mention Beverly J. Glover and Cathie Martin, "The Role of Petal Shape and Pigmentation in Pollination Success in *Antirrhinum majus*," *Heredity* 80: 778–784 No. 6, June 1998; Adrian Horridge, "Bees See Red," *Trends in Ecology and Evolution* 13 (March 1998); and A. G. Dyer, "The Color of Flowers in Spectrally Variable Illumination and Insect Pollinator Vision," *Journal of Comparative Physiology A* 183 (1998): 203–212 No. 2 August 1998.

Stephen L. Buchman and Gary Paul Nabhan, *The Forgotten Pollinators* (Washington, D.C.: Island Press, 1996), have a good discussion on the history and use of pollination syndromes.

Martha Weiss has done considerable work on the pollination behavior of butterflies. I used her research on swallowtail butterflies, taken from Martha Weiss, "Innate Color Preferences and Flexible Color Learning in the Pipevine Swallowtail," *Animal Behavior* 53 (1997): 1043–1052 no. 5.

Weiss also gave this quote in Susan Milius, "How Bright Is a Butterfly?" *Science News* 153 (April 11, 1998): "Honeybees are considered to be the intellectuals of the insect world." Earlier in the article, Milius writes that butterflies are too often "dismissed as too dumb to find their way out of a wet petunia."

Additionally, Weiss was my major source for information on how flowers change their color : Martha Weiss, "Floral Color Changes as Cues for Pollinators," *Nature* 354, November 1991; and Martha Weiss, "Floral Color Change: A Widespread Functional Convergence," *American Journal of Botany* 83, no. 2 (1995).

I should also mention Lynda F. Delph, "The Evolution of Floral Color Change Pollinator Attraction Versus Physiological Constraints in *Fuchsia Excorticata*," *Evolution* 43, no. 6 (1989).

Smelling Like a Rose

For this chapter, I am particularly indebted to two books: D. Michael Stoddart, *The Scented Ape: The Biology and Culture of Human Odour* (New York: Cambridge University Press, 1990); and Diane Ackerman, *A Natural History of the Senses* (New York: Random House, 1990). Stoddart discuss the desensitization of humans to their own smell, as well as the use of odor in human culture and perfumery. Ackerman continues and broadens that discussion.

Roman Kaiser, *The Scent of Orchids: Olfactory and Chemical Investigations* (Basel, Switzerland: Elsevier, 1993), was helpful to my understanding of how flowers produce scent, as were other books and articles. Rob Raguso, "Floral Scent Production in *Clarkia breweria*," *Plant Physiology* 116 (1998): 599–604 no. 2, gave a specific example of floral emission.

My understanding of how animals pick up scent came primarily from the Konrad Colbow, ed., *R. H. Wright Lectures on Insect Olfaction* (Burnaby, B.C. Canada, Simon Fraser University, 1989); and T. L. Payne, M. C. Birch, and C. E. J. Kennedy, eds., *Mechanisms in Insect Olfaction* (Oxford, England: Clarendon Press, 1986).

B. S. Hansson, "Olfaction in Lepidoptera," *Experientia* 51 (1995), was also helpful.

The idea of flower constancy is discussed in many books. Nickolas Waser persisted in reminding me that flower constancy is an idea still being explored. I began with his "Flower Constancy: Definition, Cause and Measurement," *American Naturalist* 127 (May 1986), as well as his "The Adaptive Nature of Floral Traits, Ideas and Evidence," in *Pollination Biology,* edited by Leslie Real (Orlando, Fl.: Academic Press, 1983).

Articles on how bees forage and smell include M. Giurfa, J. Nunez, and W. Backhaus, "Odour and Colour Information in the Foraging Choice Behavior of the Honeybee," *Journal of*

Comparative Physiology A 175 (1994): 773–779; Martin Hammer and Randolf Menzel, "Learning and Memory in the Honeybee," *Journal of Neuroscience* 15 (March 1995); and B. Gerber et al., "Honey Bees Transfer Olfactory Memories Established During Flower Visits to a Proboscis Extension Paradigm in the Laboratory," *Animal Behavior* 52, 1079–1085 no. 6, (1996).

For information on worldwide agriculture, I referred to *Collier's Encyclopedia,* s.v. "Agriculture." Vol. 21 (out of 24), New York: P.F. Collier, 1984.

Concerning the interplay of sex and food, as well as other passages, I was much helped by Elizabeth A. Bernays, ed., *Insect-Plant Interactions*, vol. 5 (Boca Raton, Fl.: CRC Press, 1994). Specifically I consulted the long and comprehensive chapter by H. Dobson, "Floral Volatiles in Insect Biology."

The tidbit that moth pheromones and elephant pheromones have similarities is from Stephen Day, "The Sweet Scent of Death," *New Scientist,* September 7, 1996, and comes from the research done by Bers Rasmussen at the Oregon Graduate Institute of Science and Technology.

Stoddart, *The Scented Ape,* discusses the experiment with women and musk and includes more information on those flower-based compounds that resemble human steroids.

Information on the giant arum and dead horse arum can be found in many books, including David Attenborough, *The Private Life of Plants: A Natural History of Plant Behavior* (Boston: Compass Press, 1995). Another source for general information is Bastiaan Meeuse and Sean Morris, *The Sex Life of Plants* (New York: Faber Publishers, NY, 1984). These books are also good sources for the variety of flowers that smell like fungi, female wasps, and the like. The story of the drag-queen flowers is also told in these books and in many others. In addition, I consulted articles such as Marlies Sazima et al., "The Perfume Flowers of *Cyphomandra* (Solanaceae): Pollination by Euglossine Bees, Bellows Mechanism, Osmophores, and Volatiles," *Plant Systematics and Evolution* 187,(1993): 51–88

Florian P. Schiestl et al., "Variation of Floral Scent Emission and Post-Pollination Changes in Individual Flowers," *Journal of Chemical Ecology* 23, no. 12 (1997), is one of many articles on this subject.

M. Gierfa, "The Repellent Scent Mark of the Honeybee *Apis mellifera ligustica* and Its Role as Communication Cue During Foraging," *Insect Society* 40 (1993), discusses the memos left by some bees.

Ackerman, *Natural History of the Senses,* describes Joy as the most expensive perfume in the world.

The Shape of Things to Come

I am grateful to my neighbors for growing passionflowers (*Passiflora incarnata*).

Peter Bernhardt, *The Rose's Kiss: A Natural History of Flowers* (Washington, D.C.: Island Press, 1999), gives thorough and wonderful passages about the shapes of flowers and the vocabulary that botanists use when talking about flowers. The quote I use from this book is in his chapter "The Pig in the Pizza."

Moore et al., *Botany*, also gives a good description of the parts of a flower.

The paragraphs on evolution were particularly difficult to write, given the complexity of the subject. I consulted a number of books, including Niles Eldredge, *Life in the Balance: Humanity and the Biodiversity Crisis* (Princeton, N.J.: Princeton University Press, 1998); Niles Eldredge, *Fossils: The Evolution and Extinction of Species* (New York: H. N. Abrams, 1991); and E. O. Wilson, *The Diversity of Life* (New York: W. W. Norton and Company, 1992). Another very readable book on the subject is David Quamman, *The Song of the Dodo* (New York: Scribner, 1996).

The latest information on monkeyflowers comes from Susan Milius, "Monkeyflowers Hint at Evolutionary Leaps," *Science News* 156 (October 16, 1999). The information on how the bills of hummingbirds evolve to fit the shape of corollas is in Ethan Temeles and Paul Ewald, "Fitting the Bill?" *Natural History* 108 (May 1999).

R. Dawkins and J. R. Krebbs, "Arms Races Between and Within Species," *Proceedings R. Society of London B* 205, 489–511 (1979), was helpful, as were Candace Galen, "Why Do Flowers Vary?" *Bioscience* 49 (August 1999), and Graham Pyke, "Optimal Foraging in Bumblebees and Co-evolution with Their Plants," *Oecologia* (Berl.) 36, 281–293, (1978).

The quote by Charles Darwin is from his *Origin of Species* (1859) and was quoted by Friedrich Barth in *Insects and Flowers*.

Sex, Sex, Sex

Lack, *Natural History of Pollination,* gives a good description and explanation of flower sex, as does Barth, *Insects and Flowers;* Bernhardt, *The Rose's Kiss;* and Moore et al., *Botany*. Karl Niklas, "What's So Special about Flowers?" *Natural History* 108 (May 1999), is helpful. I also recommend Karl Niklas, *The Evolutionary Biology of Plants* (Chicago: University of Chicago Press, 1997). In addition, I used Bob Gibbons, *The Secret Life of Plants* (Blandford, London, England, 1990).

Nickolas Waser reminded me that any discussion of natural selection has to avoid certain land mines and that any discussion on the function and usefulness of sex is still theoretical. On his urging, I consulted such articles as F. F. Green and D. L. G. Noakes, "Is a Little Bit of Sex as Good as a Lot?"

Journal of Theoretical Biology 174, 87–96 (1995); Harris Bernstein, Gregory S. Byers, and Richard Michod, "Evolution of Sexual Reproduction: Importance of DNA Repair, Complementation, and Variation," *American Naturalist* 117 (April 1981); D. G. Lloyd, "Benefits and Handicaps of Sexual Reproduction," *Evolutionary Biology* 13, 69-111 (1980); L. Nunney, "The Maintenance of Sex by Group Selection," *Evolution* 43 (1989) 245–257; and Nickolas Waser and Mary Price, "Population Structure, Frequency-Dependent Selection, and the Maintenance of Sexual Reproduction," *Evolution* 36 (1982).

I also looked at more popular articles, such as Bryant Furlow, "Flower Power," *New Scientist,* January 9, 1999.

In the Heat of the Night

A primary source of information for this chapter was Roger Seymour, a professor in the Department of Environmental Biology at the University of Adelaide, Australia, whose works include "Plants That Warm Themselves," *Scientific American,* March 1997; and "Analysis of Heat Production in a Thermogenic Arum Lily, *Philodendron selloum*, by Three Calorimetric Methods," *Thermochimica Acta* 193 (1991), 91–97. I also read Roger Seymour, George Bartholomew, and Christopher Barnhart, "Respiration and Heat Production by the Inflorescence of *Philodendron selloum* Koch," *Planta* 157 (1988); Roger Seymour and Paul Schultz-Motel, "Thermoregulating Lotus Flowers," *Nature,* September 26, 1996; Roger Seymour and Paul Schultz-Motel, "Temperature Regulation Is Not Associated with Odor Production in the Dragon Lily *(Dracunculus vulgaris)*" (poster presented at Sixteenth International Botanical Congress, St. Louis, Mo., August 1999); and Roger Seymour and Amy J. Blaylock, "Switching of the Thermostat: Thermoregulation by Eastern Skunk Cabbage (*Symplocarpus foetidus)*" (poster presented at Sixteenth International Botanical Congress, St. Louis, Mo., August 1999).

Several other articles were helpful for this chapter as well. They included Bastiaan Meeuse and Ilya Raskin, "Sexual Reproduction in the Arum Lily Family, with Emphasis on Thermogenicity," *Sexual Plant Reproduction* (1988) 1: 3–15; Gerhard Gottsberger and Ilse Silberbauer-Gottsberger, "Olfactory and Visual Attraction of *Eriscelis emarginata* (Cyclocephalini, Dynastinae) to the Inflorescences of *Philodendron selloum* (Araceae)," *Biotropica* 23, no. 1 (1993); Hanna Skubatz, William Tang, and Bastiaan Meeuse, "Oscillatory Heat Production in the Male Cones of Cycads," *Journal of Experimental Botany* 44 (February 1993); and Bastiaan Meeuse, "The Voodoo Lily," *Scientific American,* July 1966.

Dirty Tricks

Judith Bronstein, a professor of evolutionary biology, was a good source for this chapter. Her publications included Judith Bronstein, "Our Current Understanding of Mutualism," *Quarterly Review of Biology* 69 (March 1994); Judith Bronstein, John F. Addicott, and Finn Kjellberg, "Evolution of Mutualistic Life-Cycles: Yucca Moths and Fig Wasps," in *Insect Life Cycles: Genetics, Evolution, and Co-ordination,* edited by Francis Gilbert (New York: Springer-Verlag, 1990); and Judith Bronstein and Yaron Ziv, "Costs of Two Non-Mutualistic Species in a Yucca/Yucca Moth Mutualism," *Oecologia* (1997) 112: 379–385.

Other sources were Olle Pellmyr and Chad Hurth, "Evolutionary Stability of Mutualism Between Yuccas and Yucca Moth," *Nature,* November 17, 1994; M. C. Ansteet, Judith Bronstein, and M. Hossart-McKay, "Resource Allocation: A Conflict in the Fig/Fig Wasp Mutualism," *Journal of Evolu-*

tionary Biology 9, 417–428 (1996); Judith Bronstein, Didier Vernet, and Martine Hossart-McKey, "Do Wasp Figs Interfere with Each Other During Oviposition?" *Entomologia Experimentalis et Applicata* 87: 321–324 (1998); Susan Milius, "How Moths Tell if a Yucca's a Virgin," *Science News* Vol. 156 (July 3, 1999); Jerry Powell, "Interrelationships of Yuccas and Yucca Moth," *Trends in Evolution and Ecology* 7 (January 1992); and A. J. Tyrc and J. F. Addicott, "Facultative Non-Mutualistic Behavior by an 'Obligate' Mutualist: 'Cheating' by Yucca Moths," *Oecologia* (1993) 94:173–175.

Stephen Buchman and Gary Paul Nabhan, *The Forgotten Pollinators,* wonderfully describes the partnership of yuccas and yucca moths, as do other books on flowers.

The quote from Darwin is taken from his *Origin of Species.*

The quote about how cheating pollinators are more rare than cheating plants comes from Jorge Soberon Mainero and Carlos Martinez del Rio, "Cheating and Taking Advantage in Mutualistic Associations," in *The Biology of Mutualism,* edited by Douglas Boucher (New York: Oxford University Press, 1985). These authors also discuss the idea of the *aprochevado.*

Alison Brody directed me to various articles on nectar theft and robbery, including Alison Brody and Rebecca Irwin, "Nectar-Robbing Bumblebees Reduce the Fitness of *Ipomopsis aggregata* (Polemonicea)," *Ecology*, in press; Alison Brody and Rebecca Irwin, "Nectar Robbing in *Ipomopsis aggregata*: Effects on Pollinator Behavior and Plant Fitness," *Oecologia* (1998) 116: 519–527; and Alison Brody, "Effects of Pollinators, Herbivores, and Seed Predators on Flowering Phrenology," *Ecology* 78 (6) 1997 pp. 1624–1631 no. 6; as well as others.

Meeuse and Morris, *The Sex Life of Plants,* discusses forms of flower traps, deception, and mimicry. The book also describes various "murderous" arum plants and tells the story of the water lily *Nymphaea capensis,* which drowns the hapless hoverfly. This material is also covered in other books. Ethan

Temeles and Paul Ewald, "Fitting the Bill," *Natural History* 108 (May 1999), has a good sidebar on the cruelty of flowers.

The material on daisies and armyworms comes from Dennis Bueckert, "Plant Warfare," *Canadian Geographic*, July 1994.

The role of ants in pollination is discussed in Proctor, Yeo, and Lack, *The Natural History of Pollination*.

The association of politics and the scientific idea of mutualism is brought up in Douglas Boucher, "The Idea of Mutualism, Past and Future," in *The Biology of Mutualism*, edited by Douglas Boucher (New York: Oxford University Press, 1985).

The quote by the unnamed biologist is from Nickolas Waser. His ideas can be found in many articles, some already noted. Others include Nickolas Waser and Mary Price, "What Plant Ecologists Can Learn from Zoology," *Perspectives in Plant Ecology, Evolution, and Systematics* Vol. 1/2 pp. 137–150, 1998; Nickolas Waser et al., "Generalization in Pollination Systems and Why It Matters," *Ecology* 77 (June 1996); and Nickolas Waser, "Pollen Shortcomings," *Natural History* 7, no. 93 (1984).

Time

The basic material on physics and the examples of the clocks at the bottom of the tower and the twins at different places on the earth comes from Stephen W. Hawking, *A Brief History of Time: From the Big Bang to Black Holes* (New York: Bantam Books, 1988).

Material on the cereus cactus can be found in many books. I used Gary Paul Nabhan, *Desert Legends: Re-storying the Sonoran Borderlands*, with photography by Mark Klett (New York: Henry Holt and Company, 1994). Nabhan is quoted regarding the "ugly duckling" and cuddly appeal of the cereus cactus. He also mentions that he briefly thought the flowers were left-behind flashlights. I also consulted and used material from Susan Tweit, *Seasons in the Desert: A Naturalist's Notebook* (San Francisco: Chronicle Books, 1998).

More information on the history of Silver City and its parties can be found at the Silver City Museum, run by its esteemed director, Susan Berry.

Bernhardt, *The Rose's Kiss,* has many good passages on the how and why of a flower's life span.

Tweit, *Seasons in the Desert,* and Nabhan, *Desert Legends,* are both good sources for information on the century plant.

Travelin' Man

Bernhardt, *The Rose's Kiss,* has a number of good sections on pollen, as does Barth, *Insects and Flowers;* and Proctor, Yeo, and Lack, *The Natural History of Pollination.*

The quote about the "thin and particulate sheet" comes from Douglas Boucher, ed., *The Biology of Mutualism* (New York: Oxford University Press, 1985).

I also consulted S. Blackmore and I. K. Ferguson, eds., *Pollen and Spores: Form and Function* (Orlando, Fl.: Academic Press, 1985), particularly the chapter by W. Punt, "Functional Factors Influencing Pollen Form"; and Irene Till-Bottraud et al., "Selection of Pollen Morphology: A Game Theory Model," *American Naturalist* 144 (September 1994).

The information on Neanderthal burials was mainly taken from Arlette Leroi-Gourhan, "The Flowers Found in Shanidar IV, a Neanderthal Burial in Iraq," *Science* 190 (November 1975).

For the material on the murders in Germany, I consulted R. Szibor et al., "Pollen Analysis Reveals Murder Season," *Nature* 395 (October 1998). A general article is Meredith Lane et al., "Forensic Botany," *BioScience* 40 (January 1990).

The information on the Shroud of Turin was discussed in Avinoam Danin, "Traces of Ancient Flower Pollen on the Shroud of Turin: New Botanical Evidence to Date and Place the Burial Cloth of Jesus of Nazareth" (media presentation at the Sixteenth International Botanical Congress, St. Louis, Mo., August 1999); and in many newspaper articles, including Jack Katzenell, "Plant Cues Place Shroud in Holy Land," *Albuquerque Journal,* June 16, 1999.

For the discussion on buzz pollination, I read Stephen Buchman, "Buzz Pollination in Angiosperms," in *Handbook of Experimental Pollination Biology* edited by Eugene Jones and John Little (Princeton, N.J.: Princeton University Press, 1983); and Susan Milius, "Color Code Tells Bumblebees Where to Buzz," *Science News* 155 (April 3, 1999), as well as other articles.

A number of previously mentioned books and articles talk about the life of bees. I enjoyed Susan Brind Morrow, "The Hum of Bees," *Harper's Magazine,* September 1998.

The quote from the Navajo chant is from Margaret Link, ed., *The Pollen Path: A Collection of Navajo Myths* (Stanford, Calif.: Stanford University Press, 1956).

Living Together

The article "Sensitive Flower" by Andy Coghlan in *New Scientist,* September 26, 1998 summarizes recent work on how flowers "see," "smell," "touch," and "taste." Numerous other articles, such as Stephen Day, "The Sweet Smell of Death," *New Scientist*, September 7, 1996; Garry C. Whitelan and Paul E. Devlan, "Light Signaling in *Arabidopsis," Plant Physiology Biochemistry*1998 *36* (1–2) 125–133; and Paul Simons, "The Secret Feelings of Plants," *New Scientist,* October 17, 1992, discuss these subjects.

For speculation on how plants react to thunderstorms, see Stephen Young, "Growing in Electric Fields," *New Scientist,* August 32, 1997.

Autar K. Matoo and Jeffrey C. Suttle, eds., *The Plant Hormone Ethylene* (Boca Raton, Fl.: CRC Press, 1988), gives important information about this hormone. Bernhardt, *The Rose's Kiss*, also discusses the signals that start the development of a flower.

Specifically for plant intercommunication, I looked at Jan Bruin, Maurice W. Sabelis, and Marcel Dicke, "Do Plants Tap SOS Signals from Their Infested Neighbors?" *Trends in Evolution and Ecology* 10 (April 1995); Irene Sconle and Joy Bergelson, "Interplant Communication Revisited," *Ecology* 76 (December 1995); and Marcel Dicke et al., "Jasmonic Acid and Herbivory Differentially Induce Carnivore-Attracting Plant Volatiles in Lima Bean Plants," *Journal of Chemical Ecology* 25, no. 8 (1999), as well as other articles.

Proctor, Yeo, and Lack, *The Natural History of Pollination,* provides a good base for understanding communities of flowers. More about allelopathy can be found in numerous textbooks like Moore et al., *Botany,* and in articles like Gail Dutton, "Yo Buddy—Outa My Space," *American Horticulturist,* Vol. 72 March 1993; and Chang-hung Chou, "Roles of Allelopathy in Plant Biodiversity and Sustainable Agriculture," *Critical Reviews in Plant Sciences* 18, no. 5 (1999).

Dutton, "Yo Buddy—Outa My Space," discusses how plants use their roots, as does A. Tayler, J. Martin, and W. E. Seel, "Physiology of the Parasitic Association Between Maize and Witchweed *(Striga hermonthica),*" *Journal of Experimental Botany* 47, no. 301 (1996); and Charles Mann "Saving Sorghum by Foiling the Wicked Witchweed," *Science*, August 22, 1997.

James Tumlinson, W. Joe Lewis, and Louside E. M. Vet, "How Parasitic Wasps Find Their Hosts," *Scientific American,*

March 1993, and other articles discuss the relationship of some plants to wasps.

For the mimicry of *Puccinia* rust fungi, I consulted Robert Raguso and Barbara Roy, "Floral Scent Production by Puccinia Rust Fungi That Mimic Flowers," *Molecular Ecology* (1998) 7: 1127–1136; and Barbara Roy, "Floral Mimicry by a Plant Pathogen," *Nature* 362, March 1993.

Batesian mimicry and Mullerian mimicry are described in many textbooks. I also used articles such as Barbara Roy and Alex Widmer, "Floral Mimicry: A Fascinating yet Poorly Understood Phenomenon," *Trends in Plant Sciences* 4 (August 1999); James Marden, "Newton's Second Law of Butterflies," *Natural History* Vol. 1 (January 1992); Lori Oliwens, "Royal Flush," *Discover*, January 1992; and James Brown and Astrid Kodric-Brown, "Convergence, Competition, and Mimicry in a Temperate Community of Hummingbird-Pollinated Flowers," *Ecology* 60, no. 5 (1979).

Quamman, *The Song of the Dodo,* has a good discussion on some of the controversies concerning Darwin and Wallace, as well as accounts of Henry Bates's and Wallace's collecting adventures. I also read Henry Bates, *The Naturalist on the River Amazon: A Record of Adventures, Habits of Animals, and Sketches of Brazilian and Indian Life* (Dover Publications, 1975); and Mea Allan, *Darwin and His Flowers: The Key to Natural Selection* (Taplinger Press, 1977).

The Tower of Babel and the Tree of Life

A very helpful book for this chapter was William Stearn, *Botanical Latin* (Hafner Publishers, 1966), as well as portions of Moore et al., *Botany;* and Tod F. Stuessy, *Plant Taxonomy: The Systematic Evaluation of Comparative Data* (New York: Columbia University Press, 1982).

Information about Carl Linnaeus came from many sources, including Tore Frangsmyr, ed., *Linnaeus: The Man and His Work* (Science History Publications, 1994)and Bil Gilbert, "The Obscure Fame of Carl Linnaeus," *Audubon* Vol. 86 (September 1984).

For some of the newest ideas on taxonomy, I attended a number of sessions on the subject at the Sixteenth International Botanical Congress in St. Louis, Missouri, in August 1999. I also read Brent Mishler, "Getting Rid of Species," in *Species: New Interdisciplinary Essays* (Cambridge, Mass.: MIT Press, 1999); Rick Weiss, "Plant Kingdoms' New Family Tree," *Washington Post,* August 5, 1999; Susan Milius, "Should We Junk Linnaeus?" *Science News* 156 (October 23, 1999); William Stevens, "Rearranging the Branches on a New Tree of Life," *New York Times,* September 23, 1999; and Glennda Chui, "Tree of Life Proposal Divides Scientists," *Mercury News,* September 23, 1999, as quoted on the Deep Green Web page, http://ucjeps.herb.berkeley.edu, with additional keywords "bryolab" and "greenplantpage"; "Team of Two Hundred Scientists Presents New Research That Reveals Full Tree of Life for Plants" (press release prepared by International Botanical Congress, August 4, 1999); and Jeff Doyle, "DNA, Phylogeny, and the Flowering of Plant Systematics," *BioScience* 43 (June 1993).

Flowers and Dinosaurs

For some of the newest information on the evolution of green plants, including the idea that they came out of fresh and not salt water, I attended key sessions at the Sixteenth International Botanical Congress, St. Louis, Mo., August 1999. See also, for example, Kathryn Brown, "Deep Green Rewrites Evolutionary History of Plants," *Science Magazine* 285 (September 1999), listed on the Deep Green Web page.

Loren Eiscley's essay "How Flowers Changed the World" was republished as a book by the same title, with photographs (San Francisco: Sierra Club Books, 1996).

Bernhardt, *The Rose's Kiss*, recounts the story of the evolution of flowers, as does the Moore et al., *Botany*.

Else Marie Friis, William G. Chaloner, and Peter R. Crane, eds., *The Origins of Angiosperms and Their Biological Consequences* (New York: Cambridge University Press, 1987), was an important source, particularly the following chapters in the book: Else Marie Friis, William G. Chaloner, and Peter R. Crane, "Introduction to Angiosperms"; Peter R. Crane, "Vegetational Consequences"; Else Marie Friis and William Crepet, "Time and Appearance of Floral Features"; William Crepet and Else Marie Friis, "The Evolution of Insect Pollination"; and M. J. Cow et al., "Dinosaurs and Land Plants."

See also Conrad C. Labandeira, "How Old Is the Flower and the Fly?" *Science* 280 (April 3, 1998); Ge Sun et al., "In Search of the First Flower," *Science* 282 (November 27, 1998); Peter R. Crane, Else Marie Friis, and Raj Pedersen, "The Origin and Early Diversification of Angiosperms," *Nature*, March 2, 1995; Ollie Pellmyr, "Evolution of Insect Pollination and Angiosperm Diversification," *Trends in Evolution and Ecology* 7 (February 1992); and David Winship Taylor and Leo Hickey, "An Aptian Plant with Attached Leaves and Flowers," *Science* 247 (February 9, 1990).

Information on the fossil flowers in New Jersey came from William Crepet, "Early Bloomers," *Natural History* 108 (May 1999); and Carol Yoon, "In Tiny Fossils, Botanists See a Flowery World," *New York Times*, December 21, 1999.

Information about the floral landscape of dinosaurs came from sessions at the Sixteenth International Botanical Congress, St. Louis, Mo., August 1999, such as Peter R. Crane, "Plants and

Flowers from the Age of Dinosaurs: New Discoveries and Ancient Flowers" (talk given at the congress).

Many books and articles discuss the extinction of dinosaurs and the theories and controversies surrounding that event. Among others are Carl Zimmer, "When North America Burned," *Discover*, February 1997. I should also include Frank DeCourten, *The Dinosaurs of Utah* (Salt Lake City: University of Utah Press, 1998); and Tim Haines, *Walking with Dinosaurs* (BBC Worldwide, 1999).

Kirk Johnson, "Leaf Fossil Evidence for Extensive Floral Extinction at the Cretaceous-Tertiary Boundary, North Dakota, USA," *Cretaceous Research* (1992) 13, 91–117, provided specific information about the KT Boundary and the floral extinction of that time.

Regarding mass extinction and empty niches, the theory of punctuated equilibrium is well described in Eldredge, *Fossils*.

The news about *Amborella* was presented at the Sixteenth International Botanical Congress, St. Louis, Mo., August 1999, and was reported in articles like Susan Milius, "Botanists Uproot Their Old Tree of Life," *Science News* 156 (August 7, 1999).

The Seventh Extinction

Local and regional newspapers provided statistics on the effects of the heat wave in the summer of 1999. See, for example, Bob Herbert, "When Summer Turns Deadly," *New York Times,* August 8, 1999.

Several press releases from the Sixteenth International Botanical Congress, St. Louis, Mo., August 1999, gave information on extinction, for example, "World's Biodiversity Becoming Extinct at Levels Rivaling Earth's Past Mass Extinctions"; "Nearly Half of Earth's Land Has Been Transformed by Humans: Fifty Dead Zones Found in Oceans"; and "World Conservation Union (IUCN) Mobilizes International Team of Experts to Save Plant Species." Many others also gave presentations concerning extinction and human influence on that process (e.g., Peter Raven [president of the congress], "Mass Extinction of the Earth's Plant Species: Can We Prevent It?" Jane Lubchenco, "The Human Footprint on Earth: New Research"; Mike Wingfield, "Alien Invasions: Combating Aggressive Takeovers"; David Brackett, "Survival of Plant Species: A Plan of Action for the New Millennium"; and Gregory Anderson, "Threatened Islands: Storehouses of Biological Treasures"). Many other presentations at the congress dealt also with these subjects.

See also David Quamman, "Planet of Weeds," *Harper's Magazine*, October 1988. A good book on extinctions of island life, of course, is Quamman, *Song of the Dodo*.

More information on the extinction of plant species can be found in Sally Deneen, "Uprooted," *E: The Environmental Magazine* 10 (July 1999); Carol Kearns, David Inouye, and Nickolas Waser, "Endangered Mutualisms: The Conservation of Plant Pollinator Interactions," *Annual Review of Ecology Systematics* 29, 1998, 83–112; Fred Powledge, "Biodiversity at the Crossroads," *Bioscience* 48 (May 1998); and Carol Kearns and David Inouye, "Pollinators, Flowering Plants and Conservation Biology," *BioScience* 47 (May 1997), as well as other articles and sources.

What We Don't Know

In addition to interviews and discussions with Rob Raguso, I also read Robert Raguso and Mark Willis, "The Importance of Olfactory and Visual Cues in Nectar Foraging by Nocturnal Hawkmoths," *Proceedings of Third International Congress of*

Butterfly Ecology and Evolution (Chicago: University of Chicago Press, 2000); Robert Raguso and Eran Pichersky, "A Day in the Life of a Linalool Molecule: Chemical Communications in a Plant Pollinator System," *Plant Species Biology* (in press); Natalia Dudareva et al., "Floral Scent Production in *Clarkia breweri*," *Plant Physiology* 116 (1998); and Robert Raguso and Barbara Roy, "Floral Scent Production by *Puccinia* Rust Fungi That Mimic Flowers," *Molecular Ecology* 7 (1998).

Alchemy of a Blue Rose

Bernhardt's *The Rose's Kiss* has good passages on double roses and the evolution of the roses. For more information on crossbreeding, see Steve Kemper, "Ron Parker Puts the Petals on Their Mettle," *Smithsonian* 25 (August 1994). Ron Parker is the source for the quote on flower colors and nonwhite siding.

"Programs Are Launched to Analyze Impact of Bt Corn on Monarch Butterflies," *Chemical Market Report* 256 (November 1999); and "Of Corn and Butterflies," *Time* 153 (May 1999), both concern the controversy about genetically engineered corn and its potential harm to butterflies.

For more about the blue rose, see the following articles in *New Scientist*, October 31, 1998: David Concar, "Brave New Rose"; Phil Cohen, "Running Wild"; Martin Brookes and Andy Coghlan, "Live and Let Live"; and Debbie Mack, "Food for All." I also consulted Andy Coghlan, "Blooming Unnatural," *New Scientist,* May 22, 1999; Rozanne Nelson, "Not Making Scents," *Scientific American,* September 1999; Ruth Pruyne, "Green Genes," *Penn State Agriculture Magazine* (winter 1997); and Ruth Pruyne, "Shedding Light," *Breakthroughs* (magazine for alumni of the College of Natural Resources at University of California Berkeley) (summer 2000).

Jeremy Rifkin's quote is from his *Biotech Century* (Penguin Putnam, 1998).

Phytoremediation

For the use of sunflowers to clean up radiation, I read a number of articles, including Andy Coghlan, "Flower Power," *New Scientist,* December 6, 1997. Another good article on phytoremediation was Amy Adams, "Let a Thousand Flowers Bloom," *New Scientist,* December 1997. The Sixteenth International Botanical Congress, St. Louis, Mo., had numerous lectures and presentations on phytoremediation, including the media presentation by Ilya Ruskin, "Plants That Are Decontaminating the Environment."

Information on the bitter nut kola came from Maurice Iwu, "Ethnobotany: A New Plant Discovery to Cure Disease" (media presentation at Sixteenth International Botanical Congress, St. Louis, Mo., August 1999); and from "Edible Plant Stops Ebola Virus in Lab Tests" (press release of Sixteenth International Botanical Congress, St. Louis, Mo., August 1999). Several newspapers and magazines had follow-up articles on this discovery.

Information on the rosy periwinkle comes from *Systematics Agenda, 2000, Charting the Biosphere* (distributed at Sixteenth International Botanical Congress, St. Louis, Mo., August 1999).

For information on the medicinal uses of plants, I used Michael Moore, *Medicinal Plants of the Mountain West* (Museum of New Mexico Press, 1979), as well as other sources.

For material on flower essences, I consulted Clare Harvey and Amanda Cochrane, *The Encyclopedia of Flower Remedies:*

The Healing Power of Flower Essences Around the World (Thorsons, 1996); and Anne McIntyre, *Flower Power* (New York: Henry Holt, 1996).

Material on sunflowers can be found in many books and articles, including Rita Pelczar, "The Prodigal Sunflower," *American Horticulturist,* August 1993.

Index

A NOTE ON THE TYPE

This book was set in Fairfield, the first typeface from the hand of the distinguished American artist and engraver Rudolph Ruzicka (1883–1978). In its structure Fairfield displays the sober and sane qualities of the master craftsman whose talent has long been dedicated to clarity. It is the trait that accounts for the trim grace and virility, the spirited design and sensitive balance, of this original typeface.

Rudolph Ruzicka was born in Bohemia and came to America in 1894. He set up his own shop, devoted to wood engraving and printing, in New York in 1913 after a varied career working as a wood engraver, in photoengraving and bank-note printing plants, and as an art director and freelance artist. He designed and illustrated many books, and was the creator of a considerable list of individual prints—wood engravings, line engravings on copper, and aquatints.